Transforming Rural Communities in China and Beyond

Ying Zhu • Hong Lan • David A. Ness
Ke Xing • Kris Schneider
Seung-Hee Lee • Jing Ge

Transforming Rural Communities in China and Beyond

Community Entrepreneurship and Enterprises, Infrastructure Development and Investment Modes

Springer

Ying Zhu
Int'l Graduate School of Business
University of South Australia
North Terrace, Adelaide
South Australia, Australia

David A. Ness
Barbara Hardy Institute University
of South Australia
Mawson Lakes, South Australia, Australia

Kris Schneider
Faculty of Business, Economics and
Statistics
University of Vienna
Vienna, Wien, Austria

Jing Ge
Jiangsu Broadcasting Corporation
Nanjing, Jiangsu, China

Hong Lan
School Environment and Natural Resources
Renmin University of China
Beijing, China

Ke Xing
School of Engineering University
of South Australia
Mawson Lakes, South Australia, Australia

Seung-Hee Lee
Nemopartners China Consulting Group
Beijing, China

ISBN 978-3-319-35053-0 ISBN 978-3-319-11319-7 (eBook)
DOI 10.1007/978-3-319-11319-7
Springer Cham Heidelberg New York Dordrecht London

Printed on acid-free paper

Springer is part of Springer Science+Business Media (www.springer.com)

Preface

Rural communities in China and beyond are severely disadvantaged when compared with their urban counterparts, which has led to rural-urban migration in the promise of a better life, or migration of workers to cities, with concomitant social problems such as the 'children left behind'. Whilst much attention is focussed on urbanisation and the challenges of burgeoning cities, means of revitalising rural communities have received comparatively less attention. With this in mind and assisted by the Australia-China Science and Research Fund, a multidisciplinary 'group mission' was assembled to examine 'community enterprise development with connected infrastructure systems'. The authors believe this represents one of the first attempts by a multidisciplinary research team, encompassing the disciplines of social science, business, architecture and planning, engineering, and finance and economics, to assist rural communities find sustainable and self-reliant ways for their development and transformation.

The project, led by the Australian Centre for Asian Business at the University of South Australia, involved a team of researchers from universities in China, Australia and Europe who collaborated on planning, fieldwork, workshops and the development of integrated approaches. Poverty-stricken rural communities in remote mountainous areas of China were selected as the focus of investigation, in an effort to find ways to increase their prosperity, diversify jobs and improve their social livelihoods. The research concentrated on opportunities for the development of social enterprises facilitated by social entrepreneurs and community leaders, supported by infrastructure, partnerships, finance and other enabling mechanisms. Whilst acknowledging that community development also requires education, health and other services, empowering and assisting communities to operate their own enterprises are seen as fundamental to employment generation, trade and improving incomes and self-esteem.

Being consistent with the aims of the group mission, namely to build ongoing bilateral relationships between China and Australia that could lead to more intensive investigations, the research team was able to gain a general understanding of the various contexts and challenges, with opportunities for transformation to more prosperous futures, whilst recognising that the knowledge and data are incomplete and more in-depth work is required in the future.

The group mission research focussed on rural China and was later supplemented by fieldwork and discussions within Malaysia, supported by an Australia-Malaysia Institute grant. This work, which sought to build bilateral relationships between Australia and Malaysia on the benefits of information and communications technology (ICT) for rural and remote communities, provided some additional perspectives on approaches to rural community development, which are summarised towards the end of the book.

Therefore, this work is unique and important for decision makers, leaders and entrepreneurs to have access to a wider range of options from which to select appropriate modes of community transformation, especially those which are planned and implemented in a connected-up, synergistic manner. By presenting a conceptual model and systematic planning framework, it aims to stimulate new ways of thinking and to illustrate a multidisciplinary approach to the complex challenges of rural community development. By engaging with, assisting and empowering communities, this is expected to help them uncover connections and synergies between various topics, leading to innovative and breakthrough solutions. For example, interconnected community enterprises based on clean forest products, forest carbon and ecotourism can be underpinned by local infrastructure enterprises (e.g., renewable energy, water, waste management, ICT, transport) and financial mechanisms (e.g., carbon finance), all involving skills development, leadership and social entrepreneurship coupled with partnerships with the corporate sector and investors. These interconnected approaches are expected to generate increased employment and prosperity, improve social livelihoods and benefit the environment.

The book has been prepared as a result of an Australia-China group mission supported by the Australia-China Science and Research Fund. We also wish to thank the Australian Centre for Asian Business, University of South Australia, for supporting the ongoing project administration and are grateful to Tina Morganella and Vanessa Wood for their contribution to its editing and production. The ecological development union international incorporated (EDUI) was instrumental in instigating the research by making Australia-China connections. David Ness (Chair, EDUI) and Ke Xing jointly edited the book. We also appreciate the initiative and interest of Springer in approaching us to produce this book on integrated, multidisciplinary approaches. Not least, we are grateful to the various communities that we visited for assisting so willingly with our research and welcoming us so warmly.

We hope that this volume will stimulate further interest, debate and research on this important topic, and we welcome approaches for collaborative work.

Adelaide, SA, Australia — Ying Zhu
Beijing, China — Hong Lan
Mawson Lakes, SA, Australia — David A. Ness
Mawson Lakes, SA, Australia — Ke Xing
Vienna, Austria — Kris Schneider
Beijing, China — Seung-Hee Lee
Nanjing, China — Jing Ge
March 2014

Contents

List of Figures

List of Tables

Chapter 1
Introduction

1.1 Background

Rural communities across Asia and elsewhere are severely disadvantaged when compared with urban areas, to where population is attracted by the prospect of a better life. Of China's 1.3 billion people, around 500 million are located in rural areas and are generally economically and socially disadvantaged when compared with their urban counterparts. At the same time, there are high levels of poverty in many cities, with many migrants from rural areas struggling to gain employment and improved livelihoods.

China's 12th Five Year Plan places a high priority on more equitable wealth distribution, especially between rural and urban areas, to address rising inequality, modernise agriculture, create non-agriculture rural employment and service businesses, accompanied by energy efficiency and environmental protection (NDRC, 2011). To address these inequalities, China has policies to modernise agriculture to create the 'new socialist rural village', with investment in agricultural infrastructure, improved agricultural service businesses and creation of more diversified non-agricultural rural employment. At the same time, China aims to increase the urbanisation rate to 51.5 % by 2015 (China Daily, 2011), although it should be noted that 'urban' does not only include major cities, but also towns in rural areas (Aijun, 2012). Nevertheless, if rural communities and villages can enjoy lifestyles and prosperity that approximate those of their city counterparts, there may be less need for family members to seek work in distant cities—with concomitant social problems such as 'the left-behind children' (Jingzhong & Lu, 2011).

Rural communities in China have experienced rapid changes in recent years under the government's policy of 'new countryside development' (*xin nongcun jianshe*). China is experiencing an increase in social entrepreneurship led by village leaders and capable individuals with an entrepreneurial spirit. These newly established rural cooperatives share a common focus and community values with the aim of achieving common prosperity for all villagers.

Y. Zhu et al., *Transforming Rural Communities in China and Beyond*,
DOI 10.1007/978-3-319-11319-7_1

Responding to these challenges and opportunities, this book is based on the findings of research by a China-Australia 'group mission' on the topic of 'community enterprise development with connected infrastructure systems in rural China and Australia', supported by the Australia-China Science and Research Fund. With the goal of enabling stronger communities, it examines how social entrepreneurship may underpin community enterprises and cooperatives that generate employment and improve sustainable livelihoods, when supported by appropriate and inclusive infrastructure systems and enabling technologies such as renewable energy, water, waste management, information technology and built facilities, coupled with economic and financial instruments.

It documents what is transpiring in rural communities undergoing transformation, in the context of a wider society in transition, accompanied by new forces and intersections. It identifies what they are doing in innovative and cooperative ways that could be replicated in China and beyond.

1.2 Community Development: Enabling Stronger Communities

Community control of skills, knowledge, organisation, assets and understanding of local deprived communities is 'at the heart of the definition of community development' (Craig, 2007, p. 349), and the key to enabling stronger rural communities. According to Craig (2007, p. 352) citing Beazley, Griggs and Smith (2004, p. 6), "The endpoint might thus be 'less comfortable, more empowered and awkward but self-determined communities' and a process that maintains that situation".

It is important to recognise that communities have certain skills, ideas, and capacities, but these are often latent or unacknowledged (Craig, 2007, citing Taylor, 1995). Similarly, to categorise communities as 'poor' may be demeaning; although they may lack money and resources, they are often rich in ideas, culture and community spirit. As Thailand's Community Organisations Development Institute has recognised:

> While the poor may be weak in financial terms, they are particularly rich in social terms. In Thailand's communities of the poor, there is a social force which can and does already deal with most of the economic disadvantages people experience individually. (Community Organisations Development Institute, 2014)

Community development aims to support this social force and help restore, bring to the surface or unleash these skills and ideas, so communities can have control of their own destiny. Social entrepreneurs and social enterprises play a pivotal role in helping disadvantaged rural communities improve socio-economic competencies and capacities, create and sustain local employment, and become self-sufficient and self-dependent. As indicated in Fig. 1.1, the goal of building a stronger and more resilient community can be better achieved by fostering the development of social entrepreneurship and enterprises (often in the form of rural cooperatives) with innovative and effective finance models, various essential capacity enhancement

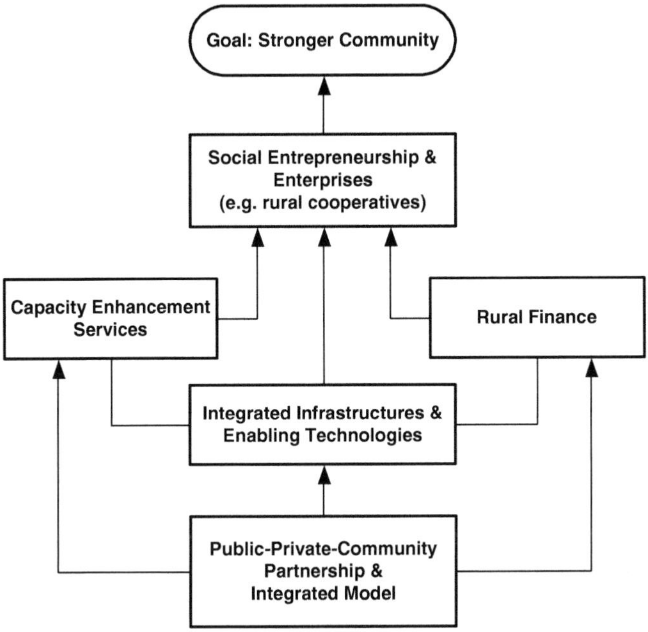

Fig. 1.1 Key elements of building a stronger community

services, appropriate infrastructures and enabling technologies, and their integration. The role of external organisations, NGOs, the corporate sector and researchers is to support communities in this process through forming productive partnerships among public, private, and community entities. Thus, the scope of the research is to create a model with connected infrastructure systems for community leadership, governance and partnership with the corporate sector, enabling investment to be attracted to rural and remote areas whilst enabling communities to exercise leadership and retain control of their own destiny.

The model would need to recognise key elements that contribute to the development of stronger communities, including: self-determination; creation of enterprises that generate employment and income; fostering of local champions, leaders and social entrepreneurs; governance mechanisms with active participation and involvement of community actors; recognising and respecting local culture, natural environment, and context; nurturing and developing local skills, ideas and knowledge; optimising the use of local resources and assets; building relationships and partnerships with external supporters, investors and advisors; ability of the community to become self-sufficient and to sustain itself over time; and enabling mechanisms such as finance, government policies and infrastructure systems.

Such elements contribute to 'human development', which comprises three basic elements: the ability of people to live long, healthy and creative lives; to acquire knowledge and skills; and to access resources for a decent standard of living (UNDP, 2013, p. 6).

Thus, using its multidisciplinary expertise, the research team sought to create an integrated model where the parts are interconnected, mutually enhancing and assist the development of stronger communities.

1.3 The Research Context

A joint research team was formed, comprising members from the University of South Australia, Renmin University of China, and Vienna University, and the first fieldwork was carried out in several rural communities in Yunnan and Zhejiang provinces in early 2012.

The rationale for selecting these two provinces was based on the consideration of comparing and contrasting rural communities with less developed and more developed regions, as well as between ethnic minority communities with *Han* (majority) communities. Yunnan is located in the south-west and is a less developed region with many remote ethnic minority communities, including the *Bulang*. Zhejiang is located on the east coast developed region with predominately *Han* communities with only one minority *She* community. The team conducted follow-up research in Baisha Country, Hainan, in November 2012, and a concluding workshop at Renmin University, Beijing.

In addition, with the assistance of funding from the Australia-Malaysia Institute, certain members of the team conducted fieldwork in Pahang Province of Malaysia related to community social uplift, the findings of which supplemented the China research (Fig. 1.2).

1.4 The Case Study Communities

*1.4.1 Mangjing Village (*Bulang *Ethnic Minority): Yunnan, Lancang County*

Mangjing village, with a land area of 89.58 km^2, is located in the southern part of Huimin Township, Lancang County which was listed as the 13th (out of 100) poverty stricken county in the world by the United Nations (2001). On the other hand, with its celebrated cultural heritage of Pu'er tea, Lancang is classified as the world's best preserved ancient tea forest in large plantation. In 2009, a Japanese expert appraised that the ancient tea-tree Jingmai Mountain resembles a world museum of tea tree plantation; thereafter, the mountain was conserved under the legislative protection of Lancang County. In addition, the Chinese Government has designated the landscape of tea forest in Mangjing as a national treasure, and its Pu'er tea was made the choice tea of the 2008 Olympic Games.

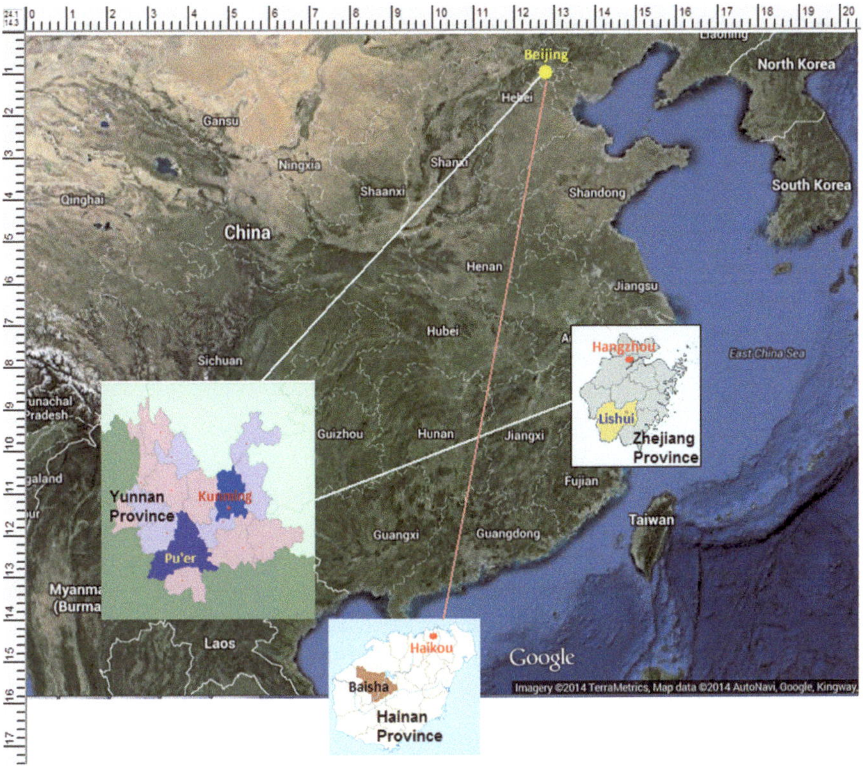

Fig. 1.2 Map showing areas studied

The highest elevation in Mangjing's mountainous territory is 1,700 m (average altitude 1,275 m) and the area has a subtropical climate. The *Bulang* ethnic population, with 586 households, accounts for about 92.1 % of the total village population of 2,645.[1] The ancient *Bulang* ethnic minority have grown tea in the area for several thousands of years and have colourful costumes and rich traditions, with the area being designated a UNESCO cultural site. The main cash crop is tea production, and Table 1.1 shows expense and income data.

Up until 2003, much terrace tea plantation was in place and chemical fertilisers were used to grow the tea tree. As a result, the quality of tea suffered from over fertilisation, land soil suffered due to mono-culture and many women's health suffered due to chemical residue contamination when picking tea leaves. The community economy was regressing, poverty was increasing and tea theft was lingering. Consequently, the village administration set up the 'Ancient tea tree protection association' in order to uphold the legendary tea culture treasured by the *Bulang* over the

[1] Source: http://news.china.com.cn/rollnews/2011-07/28/content_9213451.htm (retrieved Dec, 2012)

Table 1.1 Expense and net income by source of rural residents by region

Rural income per capita (year)	Total living expense	Total net income	Wage income	Family business income	Other income property & transfer
Nationwide (2010)	¥(4,382)	¥5,919	¥2,431	¥2,833	¥655
Yunnan (2010)	¥(3,398)	¥3,952	¥930	¥2,510	¥512
Lancang county (2011)	¥(980)	Na	Na	Na	Na
Manjing (2011)	Na	¥7,500	Na	Na	Na
Manjing (before 2004)	Na	¥515	Na	Na	Na

Source: National Statistic Bureau

last 1,800 years. In the meantime, the village sought to conserve the *Bulang* ethnic heritage and to uplift the community spirit.

In 2007, Mangjing village welcomed the establishment of the Brilliant group, which has contributed to local employment and significant export revenues. The group's tea plantation, at the area of 7.26 km^2 (11,000 'Mu'—Chinese acre), accounts for 8.10 % of the land area of Manjing village, where another part of the group's local business scope is in luxury products and services ranging from luxury resorts, boutique spas, facial products (with Pu'er tea ingredients), and also a winery and restaurant. One of the group's elite resort hotels is located on Jingmai Mountain.

By 2008, Mangjing's Pu'er tea was made the choice tea of the Olympic Games; a number of orders were placed to the village cooperative. The Brand 'Abaila' (Tea Soul) was registered to celebrate the *Bulang* heritage entering the Olympic stadium, while vigorous endeavours took place such as investments in the manufacturing building, tree saplings, training programs for production techniques and quality control alongside an aid package provided by the Ford Foundation (RMB 2000 Yuan) (Fig. 1.3).

A new mantra was created, 'One corporative, one brand', to induce five manufacturing principles which emphasised the standardisation of management, raw material collection and processing, pricing, labelling and packaging, and sales marketing. The new mantra eventually didn't survive the test of time—in 2009 many economic cooperatives and brands were formed; some brands such as '*Bulang Princess*' became quite successful while many others didn't work out.

*1.4.2 Daxi Village (*She *Ethnic Minority): Jingning Autonomous County, Zhejiang*

Jingning *She* ethnic Autonomous County is located in the southwest region of Zhejiang Province with a total area of 1,949.98 km^2 and a population of 171,867 people. Nearly 800 mountains, over 1,000 m altitude, spread over the county,

Fig. 1.3 Traditional tea culture, *Bulang*

creating high mountainous terrain and steep-cut river basins which are intersected by Ou-Jiang river streams flowing over an area of 1,725.56 km^2. Although designated as one of China's poorest counties, Jingning is abundant in minerals and is endowed with rich natural resources. The county's wetland amounts to a total area of more than 1,200 Chinese acres (0.7992 km^2), which is a sound base of ecological forestry development. In the aspect of ecological conservation, Jingning was rated number five in a nation-wide eco-environmental quality survey conducted by the China Environmental Monitoring Centre in 2005.[2] Expense and income data is shown in Table 1.2.

The community is located in a beautiful mountain setting, with forests, streams, waterfalls, walking tracks, the historic Huiming temple and a geological museum. The ancient *She* culture, exemplified by the iconic *She* multipurpose temple bridge in Daxi village (Fig. 1.4), is celebrated in a museum in Jingning City. Jingning is noted for its natural or 'wild' foods, and Huiming or 'white' tea is famous in Zhejiang Province and beyond, with other products including nutritious and valuable red rice. However, the village, region and Jingning County in general are at the developmental stage, facing the challenge to increase tourist numbers and expand enterprises while preserving the natural and cultural heritage assets.

[2] Source: http://baike.baidu.com/view/331303.htm (accessed Dec 2012).

Table 1.2 Expense and net income by source of rural residents by region

Rural income per capita (year)	Total living expense	Total net income	Wage income	Family business income	Other income property and transfer
Nationwide (2010)	¥(4,382)	¥5,919	¥2,431	¥2,833	¥655
Zhejiang (2010)	Na	¥11,303	¥5,823	¥4,307	¥1,173
Jingning (2010)	¥(5,024)	Na	Na	Na	Na
Jingning (2011)	Na	¥5,000	Na	Na	Na
Jingning (before 2004)	Na	¥600	Na	Na	Na

Source: National Statistics Bureau

Fig. 1.4 Daxi village, showing multipurpose *She* bridge

*1.4.3 Jingning Village (*She *Ethnic Minority)*

In order to overcome poverty, the Chinese Government has a policy of resettling inhabitants of more remote mountain villages—where it is difficult to provide access to services—to more urban settings, providing incentives such as 5,000 RMB per person and 60,000 RMB towards the cost of a new urban apartment. However, there is a risk that ancient *She* ethnic culture and heritage, which is so much allied to mountain settings, may be diluted or even lost. The researchers

Fig. 1.5 Jingning village (near Jingning City)

inspected a village near Jingning City (Fig. 1.5), which – whilst accessible by road and at the base of the mountain – was being closed.

1.4.4 Baisha Village, Lin'an, Zhejiang

This village comprises around 1,200 households in an area of 32 km^2 with 97 % forest cover. It is known for transforming its economy from cutting trees to one based upon forest conservation and accredited organic forest products: 'From people who cut trees to people watering trees'.

Baisha village is located in the most northern part of Linan city; it is 33 km^2 with 80 % of forest coverage. The forestry-tourism village sits in beautiful natural forest settings with intertwining creeks often times flowing alongside roads and pedestrian pathways. Before 1991, Baisha village was classified as a poverty stricken village; at that time, the arable land area was merely 1.2 % of total land area and deforestation was severe. Nowadays, Baisha village has been prized as a model village for many attributes, such as the National Civilised Village, the most attractive leisure village, new rural development model village, and common prosperity and eco-development model village.

Recognising the economic value brought about by sustainable forestry, the villagers undertook eco-restoration and forestation efforts. At the same time, they also started the micro tourism business, 'Nong Jia Le', turning their households into tourist home stays by promoting Baisha's unique ecological environment. In the midst of Taihu Lake basin tourism promotion, Baisha, renowned for its beautiful scenery, became a famous middle class income tourist spot in Zhejiang (Fig. 1.6).

Through micro-tourism, the net income has increased from RMB 7,493 in 2000 to RMB 30,000 per capita in 2011, as shown in Table 1.3. In 2012 Baisha village tourism provided about 5,000 beds from nearly 150 households running home stay family business. Over 65 % of Baisha's GDP is from tourism, although this is seasonal.

Fig. 1.6 Tourist development, Baisha village

Table 1.3 Expense and net income by source of rural residents by region

Rural income per capita (year)	Total living expense	Total net income	Wage income	Family business income	Other income property and transfer
Nationwide (2010)	¥(4,382)	¥5,919	¥2,431	¥2,833	¥655
Zhejiang (2010)	Na	¥11,303	¥5,823	¥4,307	¥1,173
Lin'an city (2010)	¥(9,339)	Na	Na	Na	Na
Baisha (2011)	Na	¥30,000	Na	Na	Na
Baisha (2000)	Na	¥7,493	Na	Na	Na

Source: National Statistics Bureau

Fig. 1.7 Baisha County, Hainan

1.4.5 Baisha County, Hainan

Baisha County is one of the poorest Counties in Hainan Province as well as in China in general, with 431 natural villages and a population of 200,000, with most of these being ethnic *Li* and *Miao* nationalities (60 %). Major agricultural commodities include tea, bamboo, fruits, and Chinese medicines. This area has rich mineral resources but, due to environmental protection (Fig. 1.7), the local government refuses to allow mining companies to develop mines in this region. The infrastructure challenges include developing waste treatment facilities in the villages in order to protect the upstream of rivers in this region.

1.5 Methodology

For the China research, we adopted the methodology of an exploratory qualitative case study analysis, which is considered appropriate for a relatively new research area and tackling research questions rather than testing hypothesis (Yin, 2008). The exploratory case study, an inductive study for refining existing theories and making complementary contributions, is also suitable for research questions related to 'how' or 'why' in an unexplored field (Edmondson & McManus, 2007). Using some published information pertaining to certain case studies of rural community development, infrastructure and finance, as well as our Chinese research partners' connections, we conducted semi-structured interviews with a number of village chiefs, individual entrepreneurs, village members, professional people assisting village development, the representatives of local People's Bank branches, and local government officials.

1.6 Structure and Content of Chapters

This chapter introduces the background to the research, its importance in the context of China and other countries, the rationale for the selection of the case study communities in rural China, and the composition of the research team. It also describes other research on the topic, the perceived gaps, and the opportunities for multidiscipline approaches—building upon the individual and group expertise of the team—to add new perspectives and develop innovative approaches and solutions.

Chapter 2 explores the key issues related to the role of social entrepreneurship and leadership in rural community development. In recent years, one of the important elements leading the changes in rural China is the role of social entrepreneurs amongst the newly established rural cooperatives managed by village leaders (*nongcun daitouren*) and capable individuals (*nengren*). These social entrepreneurs display both the entrepreneurship spirit and shared values of common prosperity for all villagers. These individuals lead and manage the rural communities with a clear social mission through the development of villagers' cooperatives as collectively-owned enterprises. Given the challenging political, social and economic environments in rural China, it is important for us to understand their role in the process of rural community development. Therefore, in this chapter, we provide six case studies of social entrepreneurs in rural communities in China. Based on these evidences, a number of implications for policy and practices are generated at the end of the chapter.

Chapter 3 analyses a green platform for a pilot transaction of China forest carbon sinks led by the Huadong Forestry Exchange in the context of a platform, demand, supply, pricing mechanism, purchase agreements and transaction products with the focus on its financing mechanism, known as China Green Carbon Fund. The chapter highlights key areas of challenges and implications that need to be addressed to continue program development of forest carbon sink projects with the support of financing mechanisms. This study provides possible implications for other provinces or municipalities at a local level in China and other developing countries preparing for, or in the process of, facilitating trading platforms for forest carbon sinks at a national level. This chapter also provides a comparison of the development of voluntary forestry carbon sinks in Yunnan and Zhejiang Provinces. It draws comparisons from four viewpoints: regional financial support, development of organisations, talent development, and exchange of experience. Finally, it concludes that capabilities of forestry resources appraisal should be improved, financial support and experience exchange strengthened, and carbon credit trading strongly promoted.

Chapter 4 focusses on carbon trade, forestry land rights, and farmers' livelihoods in rural Chinese communities. At present, China is establishing and developing a voluntary forest carbon sinks market, a process associated with forestry property transactions, as part of the domestic portfolio of greenhouse gas (GHG) emission reduction activities through market-based mechanisms. With the land property right reforms in China, the use rights of forestland were assigned to households. While forestry carbon trading is suitable for large- or middle-scale forest farms, households with a small piece of forestland find it difficult to participate in this carbon trade. Thus, China's government has combined the forest carbon trade policy and forest rights reform in order to realise the goal of climate change adoption, mitigation and poverty reduction. The implementation of these policies has had a great impact on rural communities in China. This chapter introduces the background and development process of China's forest carbon trade, analyses the impacts of forest carbon trade and forestland rights exchange on local households, based on the case studies in Guangxi and Zhejiang province, and provides some policy suggestions on how to facilitate the carbon trade and gain more benefit for local rural communities.

Chapter 5 explores alternative energy development in rural Chinese communities. Access to energy has been a critical issue for rural communities and local industry development, especially in remote mountainous regions due to inconsistent power supply, lack of modern energy services and supporting infrastructures. The rural poor are often disproportionately dependent on fuel wood and solid biomass, causing environmental degradation, reduced productivity and the eventual discount of income generating opportunities. While the United Nations declared 2012 to be the 'International Year of Sustainable Energy for All', previous case studies reveal that socio-economic growth is driven by the bundle of useful energy works and service systems rather than by simple raw energy input. The chapter provides the backdrop of world energy access and Chinese renewable energy situations, with a literature review concerning the inter-relationship of energy, poverty and development. The fieldwork in China's mountainous regions exposed practical challenges for rural development starting from basic energy service delivery to poverty alleviation. Finally, the chapter proposes to deepen the current framework for energy development in the rural context, addressing energy, poverty, and development issues with synergistic green entrepreneurship and business models to improve community livelihood and social capitals.

Chapter 6 examines and tests the proposition that stronger communities, with social entrepreneurs, enterprises and improved livelihoods and prosperity, will result from 'connected up', holistic, synergistic and inclusive planning of services and supporting infrastructure. Previously published methods for synergistic community transformation are adapted to the various contexts, beginning by understanding the challenges, unique characteristics of local culture, demographics and natural environment and resources. Various community actors may then seek to accommodate their world-views so they may work in unison towards a shared goal of transformation. Capabilities of the partners, together with local resources, assets, infrastructure and networks, may be marshalled to form 'solution oriented partnerships'. Connections and synergies may then be found between services and infrastructures so they are coordinated across sectors and are mutually reinforcing. The application of the method illustrates that more outputs in terms of community services, enterprises and jobs may be delivered with less inputs such as resources, infrastructure and cost, while conserving and respecting local culture, heritage and natural settings. Inclusive community and infrastructure planning, construction and management, employing local skills, resources and materials, may also boost local economic participation and result in more equitable, affordable and innovative solutions.

Chapter 7 analyses information and communications technology (ICT) based service innovations for supporting rural community enterprises. Despite much research and practical effort having been devoted to ICT-enabled services for rural development, a major limitation in such rural applications is the lack of a clear and systematic analysis to capture the characteristics of rural ICTs and to guide the development of collaborative partnership and systems solutions for ICT-based service innovation. Through a detailed analysis of an ICT-enabled rural financial service case in Lishui, China, it is identified that ICT-enabled rural services demonstrate a strong element of multi-stakeholder collaborative innovation. Rural ICT solutions are developed as a result of synergy-based partnership among different organisations with shared goals, interests, capabilities and resources. In addition, ICTs are not only outcomes of collaboration, but also the platform to facilitate collaborative service innovation and to help integration of different services for holistic solutions. Furthermore, ICT devices, applications, and infrastructures are embodiment or carriers of rural services. Such characteristics of ICT-enabled service innovation are essentially in line with the Product-Service System (PSS) concept and paradigm. Based on the case analysis, it is argued that a synergy-based PSS model for community transformation can be adopted to provide a systematic process to guide the rural ICTs and ICT services development.

Chapter 8 highlights the key elements for stronger rural communities, drawing together the themes and proposals from the preceding chapters and constructing an integrated model. With stronger communities at its core, the model recognises the key role of social enterprises in job creation and community transformation, supported by various actors working together, combining their capabilities and sharing their assets in a synergistic, effective and efficient manner. Set within the community's political, cultural, socio-economic and environmental context, the model also highlights the importance of community empowerment, equitable development and cross-organisation governance. The chapter also sets out a direction for further research, including on the ground application and testing of the integrated model, and the extension of the approach to other communities in China and beyond.

Special acknowledgement This book has been prepared as a result of an Australia-China 'Group Mission' supported by the Australia-China Science and Research Fund.

References

Aijun, Q. (2012, March). How to understand the urbanisation rate in China? *Cities Development Initiative for Asia.*

China Daily. (2011, March 5). China expects 51.5 % urbanisation rate by 2015. *Xinhua.*

Community Organisations Development Institute. (2014). *Baan Mankong collective housing,* Thailand. Retrieved January 12, 2014, from http://www.codi.or.th/housing/aboutBaanmankong.html

Craig, G. (2007). Community capacity-building: Something old, something new…? *Critical Social Policy, 27*, 335–358.

Edmondson, A. C., & McManus, S. E. (2007). Methodological fit in management field research. *Academy of Management Review, 32*(4), 1155–1179.

Jingzhong, Y., & Lu, P. (2011). Differentiated childhoods: Impacts of rural labour migration on left-behind children. *The Journal of Peasant Studies, 38*(2), 355–377.

NDRC. (2011). *China's twelfth five year plan (2011–2015), National Development and Reform Commission* (European Delegation to China, Trans.). Retrieved February 28, 2014, from http://www.britishchamber.cn/content/chinas-twelfth-five-year-plan-2011-2015-full-english-version

UNDP. (2013). *China human development report 2013. Sustainable and liveable cities: Toward ecological civilisation.* United Nations Development Program, China and Institute for Urban and Environmental Studies, Chinese Academy of Social Sciences, China Publishing Group Corporation. Retrieved http://www.undp.org/content/dam/china/docs/Publications/UNDP-CH_2013%20NHDR_EN.pdf

Yin, R. (2008). *Case study research: Design and methods* (4th ed.). Beverly Hills, CA: Sage.

Chapter 2
Rural Community Development and the Role of Social Entrepreneurs

2.1 Introduction

This chapter examines the key role of social entrepreneurs and social enterprises in building stronger communities. Through the development of village cooperatives as collectively-owned enterprises, these individuals lead and manage the rural communities with a clear social mission - an alternative business model to business operations under market capitalism that focuses on self-wealth and its expansion. Six case studies of social entrepreneurs in rural communities in China are discussed, which show there are multiple key characteristics of their leadership role in uniting rural community members to achieve common prosperity. A number of implications for policy and practice are generated at the end of the chapter.

2.2 Social Entrepreneurs in Rural Communities in China

Since economic reform started in the early 1980s, there have been two models of economic development in rural areas of China, namely a top-down systematic reform led by the governments and a bottom-up innovative change of rural production and business operation initiated by farmers themselves (Liu, 2004). The bottom-up oriented changes have created new systems as well as new production and business opportunities that are led by individuals or groups of individuals labelled as 'rural capable people', who are able to identify opportunities and take risks to lead other villagers to achieve common prosperity. The planning system in general and the commune system in the countryside in particular have gradually been replaced by a market-oriented economic system and individual household responsibility system as an important part of China's rural economic reform. Under the new policy, farmers have been able to obtain greater autonomy and responsibility for running their own production and business, and there has been an increasing desire

Y. Zhu et al., *Transforming Rural Communities in China and Beyond*,
DOI 10.1007/978-3-319-11319-7_2

amongst these farmers to create wealth for the individual household as well as their village communities. A number of social entrepreneurs were brave enough to break the 'old fence' and develop new enterprises and small household business networks. As a result, a rapid increase of household income amongst these people in the first couple of years of their new ventures influenced other villagers to follow them and to do the same thing under these social entrepreneurs' support. This bottom-up development has built a foundation of new initiatives of agri-business such as township and village industry enterprises, household holiday houses for tourism and family-owned agri-production and trade.

Four types of rural social entrepreneurs, known in Chinese as *nongcun daitouren*, can be identified (Wan, 2007; Wu, Yu, & Huang, 2008; Zeng, Li, & Yang, 2005). The first type is the existing village leader, elected by villagers and in that position for many years, maintaining a good reputation within the village, and well-connected outside of the village. They lead other villagers to set up new enterprises in the form of village cooperatives for developing agri-business. Most of these leaders have good will for achieving common prosperity amongst the community and their role is to lead everyone to achieve their goals. The second type is the new business person who has run their own business successfully within the village for many years and has been seen as a good example for other villagers to follow. Most of them are capable entrepreneurs with a wide range of skills and social networks. They normally use their own business as the foundation to form new farmers' cooperatives that include other villagers participating in the business operation and sharing the benefits. The third type is the migrant worker returning to the home village with many years of industrial and business experience accumulated through working in cities. On their return, they bring new business opportunities, management skills and their social networks with them. New business cooperatives can be developed under their initiatives. The fourth type is the professional person (i.e. engineers and professors) outside rural communities that work together with rural communities and develop special products and business with particular know-how and techniques that benefit the rural communities. Their intention is to utilise their technical skills and knowledge to support rural communities to develop unique production techniques and business.

Generally speaking, the rural social entrepreneurs have some extraordinary capabilities that other villagers may lack. Huang and Cui (2009) summarise three key capabilities, namely self-learning and obtaining knowledge, applying knowledge to practice, and innovation and creativity. These capabilities enable them to accumulate, increase and realise their human capital. As for the rural social entrepreneurs, their basic knowledge and skills were obtained from their formal school education, training provided by the local government and other training institutes, and technical development obtained from their own professional experience. However, the most important factor for their entrepreneurship capability development is their previous experience. The concept of 'learning on the job' is very important for them to accumulate knowledge and skills and then learn how to apply them to practice. This is particularly relevant to the situation in China given the market-oriented business environment has only developed in the last three decades. In addition, in recent years, they have built up social networks for obtaining information and knowledge related to production, business opportunities and marketing (Li, Schulze, & Li, 2009). These networks include a wide range of business and professional people with different expertise, such as people doing the same type of business, sales agents, customers, relevant government officials, technical and professional people, and even competitors. Learning and development for the rural social entrepreneur is a process of continuous revitalisation and improvement.

Regarding the unique characteristics of the rural social entrepreneurs, Liu (2004) points out that they are ‘the traditional rule/system breakers’ and ‘new rule/system creators’. Generally speaking, the cultural norm and way of life in the countryside in China is overwhelmingly conservative and many traditional rules and systems place constraints on innovative ideas and actions. Farmers are accustomed to the traditional way of doing things and challenging the existing rules and systems is a difficult task. However, economic reform requires people in rural communities to take the lead by changing and breaking the old rules and ways of doing things. Rural social entrepreneurs face many challenges when adopting new approaches and encounter powerful resistance during the process of business development. Different stages of market transition provide different opportunities as well as constraints (Yang & Li, 2008). Therefore, these rural social entrepreneurs must be able to overcome difficulties with the quality of self-determination, proactive thinking and action, self-control and persistence, and be able to implement their strategy effectively (Li, Young, & Tang, 2012). These unique characteristics enable the rural social entrepreneur to survive during the process of reform and transformation. The following sections provide a detailed record of six social entrepreneurs’ experiences of developing community-based cooperatives in a number of villages in Yunnan and Zhejiang Provinces.

2.3 Case Studies

In our fieldwork, we identified six leading social entrepreneurs in our sample rural communities. When interviewing these rural social entrepreneurs, we observed many similarities but also some differences related to their own background, gender differences and past experiences. Given all of these villages are located in mountain regions with historically poor backgrounds, these rural leaders have a strong determination to change the poor conditions and lead their communities to achieve a better way of life. This is a very common goal amongst all of our samples. However, the way to achieve this goal is different for each with many difficulties and even failures. Therefore, in this section, we plan to present the cases by firstly providing a brief background of these individuals and then analysing these cases by illustrating the process of social entrepreneurship in challenging and difficult environments. The eventual goal is to explore the role of social entrepreneurs in rural community development in terms of achieving the social missions of sustainability and common prosperity in a broader competitive environment.

2.3.1 Social Entrepreneur (A)

The social entrepreneur (A) is a village chief at Mangjing village, a remote *Bulang* minority community in Lancang County, Yunnan Province. The village chief has held this position for many years since rural reform began in the 1980s. He experienced the transformation from 'commune system' to 'individual household responsibility system' then to 'voluntary cooperative system'. Furthermore, he has been the driver of change by implementing relevant government policies, confronting challenges and difficulties, and being actively engaged with both internal and external stakeholders. Mangjing village has been a model village of common prosperity and eco-development for many years and was nominated as the '2011 most attractive leisure village in China'.

However, the village's development has not always been successful and many difficulties have been experienced—particularly in the early years of rural reform. In the early years, thinking and action were driven by short-term making quick cash income without much consideration given to long-term well-being and sustainability. For example, the village had many ancient tea trees that were estimated to be over 1,300 years old. These ancient tea trees produced high quality tea but in limited quantities. In the 1990s, individual households began cutting down the ancient tea trees and replacing them with young tea plants in order to achieve a higher quantity of tea production. They used intensive monoculture tea plantation practices called 'tea gardens on the terraces' but there was little biodiversity. This new way of establishing tea plantations relied heavily on the input of fertiliser and other chemical products for pest control. The initial results saw a greater output of tea production in terms of quantity, but the quality of the tea suffered and did not compare with the quality of the ancient tea. Moreover, after several years of this type of intensive production the soil and surrounding environment was adversely impacted. Eventually, the overall tea production in terms of quantity and quality declined due to soil degradation caused by intensive monoculture practices and the amount of chemical products used.

It became obvious to the villagers that this type of tea production was not sustainable and household income could not be maintained in the long-term. The village chief led the change by inviting professional people from Yunnan Agricultural University to advise them on changing the way they cultivated tea. They started protecting and revivifying the ancient tea gardens, and then adjusted the 'new tea gardens on the terraces' by reducing the level of cropping intensity and adopting traditional *Bulang* tea plantation methods. These changes resulted in a much lower input of chemical products being required and increased the biodiversity of the area by mixing different plants as well as sustaining old and new tea trees. The forest canopies improved the ecology of the tea gardens and enabled the local people to harness the ecosystem services of rich biodiversity to support their tea production. Under the leadership of (A), the village set up the Ancient Tea Gardens Protection Committee in 2005 and Mangjing Ancient Tea Cooperative in 2006. Subsequently, the village has developed their own tea brand 'Arbaila', namely 'tea spirit' in the *Bulang* language, and this tea was nominated as an officially selected tea for the 2008 Beijing Olympic Games.

As the village chief and leader of the village cooperative, (A) defined his role very clearly. He said that he was always thinking of the well-being of the entire village first, and that he needed to provide a good example to lead other village members by putting his own money into the cooperative as a deposit, mortgaging his own forest land to borrow money from the bank, converting his house to develop household tourism, and tirelessly developing his business networks for the village business. His philosophy was, "As a leader, you have to lead with a good example and other village members will follow. Given that our village people are very conservative, you have to demonstrate that you can achieve financial success by conducting agri-business in a way that maintains sustainability—and then they will follow. Actually, a leader has to take risks from time to time and not every business deal will make profit".

In addition, (A) was very conscious about protecting the tea's brand name. He established an up-rule governed by five unified principles: unified management, unified tea collection, unified price, unified brand, and unified packaging and sales. He was actively engaged with external stakeholders such as Yunnan Social Science Academy and Provincial Foreign Investment Centre and obtained funding support from Ford Foundation with RMB 200,000 to build a tea processing factory and grow more tea saplings. He worked together with the County Government to develop new roads, and ensure that there was electricity and water supply to the village. He said, "Sustainable development needs to rely on government support first, and secondly on the cooperation of village members. If we all have the consciousness of cultural and environmental protection, in particular in the areas of ecotourism and preserving ancient tea gardens, then our *Bulang* culture and village livelihood can survive in the long-term".

Under the leadership of (A), the per capita annual income increased from RMB 515 before 2004 to the current level of around RMB 7,500—an increase of more than 14 times. This resulted in most young people staying in the village as there was ample employment available and they did not need to seek jobs in the cities. Our impression of the village was that it is very clean and looks much like a public park. The *Bulang* culture has been well preserved and a very good school has been provided with donations from external stakeholders. A training centre, organized by the *Bulang* Prince and funded by UNESCO, has also been established for passing on traditional cultures.

However, ongoing problems and new challenges are confronting the village and its leader. For example, in 2009, the government promoted more cooperatives to be established at the village level with the policy that any five households could form a cooperative. Since then, more than 20 small cooperatives have been established at Mangjing community. These small cooperatives were not competitive enough in the markets and they also dispersed the brand name. Therefore, "another round of unified activities amongst those small cooperatives became necessary". (A) also mentioned other challenges, including insufficient infrastructure and financial support, skills and knowledge development, and attracting capable people from outside the village. We felt that (A) is a very capable leader with good communication and network building skills and a clear vision for achieving the goals of sustainable development and common prosperity as his social mission.

2.3.2 Social Entrepreneur (B)

The social entrepreneur (B) is a village capable person at the same village as (A). She is a female leader of a village-based cooperative and a village women's group. She organised the village women to form a *Bulang* minority singing and dancing group and is engaged in producing *Bulang* costumes, arts and crafts products and receiving tourists at her household hotel. As a capable person in the village, she has experience from working in Lancang City as a school teacher and is self-trained with tea production capability. Her mother passed on *Bulang* traditional cultural knowledge to her such as singing and dancing, and how to create products using cultural traditions for the tourist market.

During the interview, (B) mentioned that she quit her job as a school teacher in Lancang City and returned to her home village many years ago in order to look after her family business and her parents. After returning to the village, she wanted to create a more enjoyable living environment for the village people in the same way that city people enjoyed. She volunteered to be a leader, brought together a women's singing and dancing group and also helped the village school to develop social activities for the children. She felt that preserving the *Bulang* traditional culture was a matter of urgency so she organized a training class and invited her mother to instruct and teach village women how to make the *Bulang* costumes, arts and crafts products. In order to be familiar with tea production, she went to the tea mountains and tea production factories to learn the tea production process from experienced people. Then, when a new policy was implemented in 2009 to promote village-based cooperatives, (B) used her family tea production capacity as a platform to form a tea production cooperative. She worked with 20 households to establish the cooperative and used her family forest and other members' forest use-rights to mortgage a bank loan for developing the cooperative. She then shared the revenue amongst the 20 household participants. (B) established a good example for other village women to follow and she played a nurturing and supportive role to village women and children as she was very keen to preserve the traditional culture for current and future generations. Both (A) and (B) identified similar issues and challenges around the needs of future development, in particular around the issues of brand name development and marketing.

2.3.3 Social Entrepreneur (C)

The social entrepreneur (C) is a village chief at Daxi village which is a relatively remote *She* minority community in Jingning County, Zhejiang Province. He is a relatively young village leader (in his early 40s) compared with (A) and he was previously a school teacher and administrator before becoming a village chief. He trained at the Party Cadre College and during this time accumulated a wide range of knowledge and networks. (C) was the primary driver for the establishment of a village-based cooperative and he has remained watchful of the changes in the markets, which has allowed him to make adjustments to the agri-business when necessary.

Jingning County in general and Daxi village in particular have always been one the poorest regions within Zhejiang Province, largely due to the remote location in the mountains. For many years up until 2004 (see Table 2.1) the annual per capita income level was below RMB 600. The old agricultural production was narrowly focussed on rice production, which resulted in villager's cutting down trees to build 'rice terraces' on the mountains. However, the production output was very low and consequently the villagers' income remained at a very low level. In 2006, the first village cooperative was established by diversifying the agri-business into higher value-added production and ecotourism. The new products included wild rice stem and red rice which is a more nutritious food that fetches a higher price. Also, the villager's replanted bamboo and tea for preserving forests and producing high price commodities from them. As (C) claimed during the interview, "We are in a less developed region, we need to protect our natural resources, namely our forests and surrounding environments first, then develop our local economy and business according to our natural advantages so we can achieve sustainability and not make the same old mistakes". A good example was the discovery of the value of red rice. Red rice has the nickname 'wild boar fear' because it was thought that the wild boar fears the harsh shell of the red rice. Local farmers utilise the red rice plant as a fence to protect the rice field from wild boar intrusions. However, in recent years, red rice was discovered to be a highly nutritious food with the function of reducing blood sugar levels. It suddenly became a valuable commodity in the city and the price of red rice increased. Therefore, the production of red rice has become one of the major agri-businesses in recent years.

(C) was very clear about capitalising on the uniqueness of the village and its products. He said:

> Shanghai has a lot of high rise buildings, but we have many high mountains. We have *She* culture, clean water, air and a pristine environment which Shanghai does not have. We need to protect the natural beauty of our village and our cultural traditions, including preserving the old architecture. We do not need to follow the path of the developed region of fast growth. We can develop in a balanced way, by improving public services and facilities, combining our culture and ethnicity with ecological and economic development in a sustainable way.

Cooperatives were formed based on a loosely organised association among villagers with the model of 'large household leading and small household following'. As (C) claimed, "Farmers generally do not want to take risks and they are not able to afford any risk. Therefore, leaders and village capable people have to lead and take the risks. When we achieve a better outcome, then other farmers will follow and receive the benefits". The villagers formed different cooperatives specializing in a particular commodity, such as wild rice stem cooperative, red rice cooperative, bamboo cooperative and tea cooperative. Each cooperative elected its leaders and cooperative board members and those regarded as village capable people. Once again, our impression of the village was that it was very clean and well organised, much like a public park. The mountains were green with all kinds of trees, but wild rice stem, bamboo forest and tea dominated a large proportion of the landscape. The annual per capita income rose from RMB 600 before 2004 to RMB 5,000 in 2011, which is more than an eightfold increase. During the interview, (C) presented himself as a well-educated and well-informed village chief with good connections to governments and business networks outside the village. He is a very ambitious young leader with a clear vision and goal for achieving a better village life for everyone.

2.3.4 Social Entrepreneur (D)

The social entrepreneur (D) is another village chief at Baisha village which is located in a more prosperous region of Linan County in Zhejiang Province, not very far from Hangzhou, the capital city of Zhejiang, and Shanghai when compared with the other remote villages. Baisha village has been a model village of common prosperity and eco-development for many years and in 2011 it was nominated as 'the most attractive leisure village in China'. The village chief has held this position for many years. Previously he was a member of the army and also worked in the city and has a wide range of social contacts. (D) has led the village's transformation from cutting down trees and selling wood products in the earlier years to the now much richer condition of preserving trees and developing a diverse agri-business. In recent years, villagers have tried to fully capitalise on the natural advantages of the region by engaging agri-business in the plantation and production of bamboo, tea and nuts in the mountains and have developed village-based eco-tourism and hotels.

The village chief (D) told us that Baisha village has experienced three stages of transformation: the first stage was before 1990 when the major production activities were cutting down trees and producing rice on the mountain terraces; the second stage was between 1990 and 1997 when the major activities were selling natural mountain products to the markets in order to make quick money; and the third stage was after 1997 when people realised that the old way of doing things was not sustainable. They changed their way of thinking about development patterns and began to preserve the mountains by planting trees, bamboo and tea. They began to establish biodiversity and improve the surrounding environments to attract ecotourism. These positive changes came about because of the negative lessons they learnt from the early years of development when frequent floods and landslides occurred due to forest damage.

During the interview, (D) stated, "Baisha village is located at the upper stream of Tai Lake. Therefore, it is important to preserve the mountains by planting trees, bamboo and other plantations in order to keep the water clean and prevent landslides during the rainy season. Our choice for village development must consider the natural conditions first and then select suitable development models. Ecotourism is one of the options we have". Nowadays, every village household has built a home-based hotel and some are capable individuals who have also set up tourist companies to take city people from Shanghai and Hangzhou to the village for a holiday. "We have more than 200,000 tourists visit here each year. Even the famous NBA basketball player Yao Ming visited here for his wedding photo shoot. We have a unique natural environment with beautiful mountains, clean water, clean air and delicious food such as fresh baby bamboo, prawns and fish, and dry nuts—city people like these".

We asked questions about the current challenges and problems. (D) replied:

> One notable challenge is how to build our brand name and maintain a good reputation. As for the development of eco-tourism, we also need to improve our infrastructure facilities, such as ICT for communication, waste treatment and other services such as health care and medical facilities for village people and visitors. Bank facilities have improved a lot and our local shops are linked with banks and use EFTPOS devices and ATMs. Another problem is the inconsistency of our architecture. Given individual households are building their own hotel, different shapes, colours and designs appear in the one village and it looks terrible. We have to be cautious about preserving the traditional style of architecture and use local materials. Meanwhile building design must be consistent with the natural environment of the village. We also need to attract capable people from outside to help us to build a professional management team and develop our agri-business and ecotourism effectively.

The Baisha village has achieved a lot in the past 10 years with a dramatic increase in annual per capita income from RMB 3,804 in 1997 to RMB 30,000 in 2011. Our impression of the village is that it is a very prosperous one with beautiful green mountains (97 % forest coverage) frequented by many tourists. One obvious problem is that so many new hotels are being built with different colours and styles that do not match the local village character and are not in harmony with the natural environment. As (D) indicated, these problems need to be addressed collectively in order to preserve the Baisha traditional culture. (D) gave us the impression that he is a visionary with clear goals and well-connected networks. A social mission is clearly indicated regarding sustainable development, improving income among villagers and preserving village culture and mountain environments for current and future generations.

2.3.5 *Social Entrepreneur (E)*

The social entrepreneur (E) is a village capable person who lives in the same village of (D). Several years ago before returning to the village, he was a member of the army and later worked in big cities like Shanghai and Hangzhou. He has built a good business network in these cities and his work and life experience and networks have helped him considerably in terms of developing village-based tourism and hotels as well as selling mountain products such as bamboo, tea and dry nuts into the city markets.

(E) took us to visit his cooperative-owned hotels, one old and one new, with a capacity to receive 400 visitors. He explained to us that after he graduated from high school, he joined the army for several years. After working in the cities he returned to his village and became involved in the development of village-based ecotourism. In order to gain more professional knowledge, the village committee sent him to Zhejiang University to attend formal training courses in ecotourism development. On his return, he worked together with several villagers to set up their own cooperative on ecotourism. He used his own savings and also borrowed money from the bank by mortgaging his forest and mountain land. Working together, villagers built the first hotel and also developed a mountain-based agri-business producing tea, bamboo and dry nuts. The business became very successful so more money was invested to build a second hotel where 18 workers are now employed. They also set up their own transportation teams for picking up tourists from Shanghai and Hangzhou and delivering the agricultural products to the wholesale markets. In comparison with the average annual income, the cooperative had a much higher income level of RMB 100,000 per capita. When we asked him about the key capability for rural social entrepreneurs to be successful, he highlighted "taking risk, effective judgment, flexibility, knowledge and techniques, and sincere communications with others—these are all very important". Regarding the most difficult barrier, he noted that there was a lack of monetary support from financial institutions, given that most farmers do not have many assets to use as a deposit. They had to be creative by using the customers' money in Shanghai and Hangzhou to build the second hotel and keep some nice rooms for these customers permanently in return.

Regarding the challenges, he pointed to several key areas:

> … lack of education infrastructure in the village and the closest school is 15 km away; a need to build more waste treatment facilities for maintaining eco-sustainability; preservation of traditional architecture by using local materials and consistent design practices that harmonise with the mountain environment: developing and maintaining a good brand name and improving the quality of services offered. In order to tackle these challenges effectively, we need to be able to attract more capable people from outside the village to help us develop our community in a professional way'.

As a capable individual, (D) demonstrates that he is capable of rural development renewal not only for himself and family, but also for other village members' well-being—it is apparent that he is on a clear path to achieving his social mission.

2.3.6 *Social Entrepreneur (F)*

The final social entrepreneur (F) is a retired senior engineer from Linan Forest Bureau. Before and after his retirement, he has assisted Baisha village to develop mountain-based agri-business through bamboo forest protection, plantation and development by using innovative and technological methods. After retiring from the Forest Bureau, he set up his own research and development centre and also organised a village-based bamboo business cooperative with 80 % of village households joining as members. He brought new technology and know-how into the village and shared this with village members. He also trained village leaders, capable individuals and his family members to develop new business ventures and maintain the balance between economic development and ecological sustainability.

From 1991, (F) started to bring experts from the Chinese Academy of Forestry, Zhejiang Forest College, and Linan Forest Bureau to implement a series of international cooperation projects into Baisha village. They introduced new ideas for sustainable development through effective use of forestry resources at the mountain village. This led to a change in villagers' behaviour from cutting down trees to protecting the forest. The establishment of community cooperatives increased the value of understory resources. Currently, 73 % of villagers' income comes from non-wood forest products, such as the wild baby bamboo, wild tea, pecan, and mountain vegetables. (F) led the villagers to find alternative ways to make a better life without sacrificing the natural environment.

During the interview, (F) explained the processes that villagers followed, to transform the rock-mountains into bamboo forests. It took 4 years, from the first year of breaking the large rock into small pieces and mixing it in with good soil, and then planting bamboo on it. In the second year the natural transformation from small rock pieces into soil occurred, and in the third year there was an overall increase of green bamboo on the mountains. In the fourth year products were able to be harvested to produce a cash income. During the 4-year cycle, other mountain products such as wild tea, dry nuts and vegetables were produced simultaneously. The return of cash investment (not including labour input) was more than 200 %, and most of the output was associated with baby bamboo products. After several transformative cycles the initial barren rock-mountains turned into green forests with a focus on bio-diversity and with an output of multiple mountain products.

Regarding the formation of the bamboo cooperative, (F) explained the processes: the individual household provided a cash investment (from RMB 100 to 15,000), then a new cooperative was formed with more than 100 households with a collective capital of more than RMB 200,000. Each year income returns doubled. The first cooperative project involved a round of collecting products such as bamboo, tea, nuts and so on. Individual households sold their products to the cooperative at the market price and received cash. The second one was through an end of year dividend paid out based on the individual households' initial investment. All the products have the cooperative's own brand. Given the products come from a pristine natural environment, they are very popular in the big cities like Shanghai and Hangzhou. In addition, the location of Baisha village is not far away from these cities, which has a twofold benefit of low transportation costs and ensuring the freshness of the product. These benefits translate into Baisha village products being more competitive and more attractive to consumers than other products.

(F) has been using the research centre as a training base for people coming from other villages and provinces. They have also been engaged with many international institutions by cooperating with and providing a new source of export market (such as bamboo seedlings to Africa and the Middle-East). When we discussed these initiatives with village members, they all praised (F's) significant contributions to the village and his leadership, use of knowledge, up-to-date technology and social networks. His passion for developing a better village life and ecological development of rural mountains has generated new hopes for village people as well as sustaining the local environment. (F) has made a key contribution to transforming processes in his village communities' sustainable development.

2.4 Findings and Implications

This study is based on in-depth interviews with several rural social entrepreneurs in multiple villages in Yunnan and Zhejiang Provinces in China. It aimed to explore the key issues related to the leadership role of social entrepreneurs in rural communities' development. Based on studying six cases of rural social entrepreneurs, the findings identify that there are multiple key characteristics of their leadership role in the process of establishing and developing rural cooperatives that could lead and unite rural communities' members to achieve common prosperity, as summarised in Table 2.1. In this final section, we identify the key findings and implications for rural community development and the role of social entrepreneurs.

The first finding is that all the cases have special traits and special leadership skills with some common elements as well as differences. The major common aspects of special traits include the consideration of collective well-being and achieving common goals, leading by good example, risk-taking, self-sacrifice, and acting for a social mission. The major differences appear between the village chief and individual capable person. As for the village chiefs, holistic consideration for decision-making and motivating other village members are significant given the role of the village chief. On the other hand, the individual capable persons play a more important role in volunteering to work for the benefit of the village. They are more concerned about particular issues that relate to children's education and women's welfare. In addition, female entrepreneurs play an important nurturing role to manage issues such as preserving traditional culture, organising women and children's activities and caring for the elderly.

In terms of special skills, the village chiefs are normally astute in the areas of leadership, are experienced with managing village affairs, are effective communicators when engaged with internal and external stakeholders, and are strategic thinkers and generally multi-skilled. As for the individual capable persons, they normally have special skills associated with their previous work experience such as being a school teacher (such as (B)), or having been an army member, and having worked in the city (such as (E)), and being a professional person (such as (F)). Some common elements also exist among them, such as they are all very capable of network building and vigilant when capturing business opportunities. They are all eager to learn new things and share their knowledge amongst fellow villagers.

The second finding is that different stages of business and venture development require different strategies and actions adopted by the rural social entrepreneurs. In

Table 2.1 The leadership role of six rural social entrepreneurs

	A	B	C	D	E	F
Leadership skills & knowledge	Leadership skills and experience of managing village affairs; effective communications with internal and external stakeholders; strategic thinking and multi-skilling	Knowledge of *Bulang* culture (singing & dancing, costume, arts and crafts production); tea production process; knowledge as a school teacher; experience of leading village women's group	Leadership skills and experience of managing village and education system; skills of network building, problem identification and prevention skills; capable on-going learning and adaptation skills	Leadership skills and experience of managing village affairs; proactive and effective engagement skills with internal and external stakeholders, in particular with professional institutions and experts on agri-business and production	Experiences of working and running small business outside village; entrepreneurship skills of developing new business; network building skills with internal and external stakeholders	Professional knowledge and expertise on bamboo plantation, production, diversification, and marketing; leadership skills of managing institutions and facilitating rural community development; entrepreneurship skills of developing new business; network building skills with multiple stakeholders
Leadership characteristics	Considering collective well-being; leading with good example; risk-taking and sacrifice; flexible and persistent to achieving common goals	Leading by good example, volunteering for village activities; helping other villagers to achieve common prosperity	Efficient decision-making; flexible and adaptable; risk-taking; holistic consideration for decision-making; clear purpose for achieving common goals	Leading by good example; flexible and adaptable; risk-taking, motivating other capable villagers to lead the changes; clear purpose for achieving common goals	Volunteering for village activities as village committee member; leading by good example; knowledge accumulation and sharing; risk-taking; helping other villagers to achieve common prosperity	Leading by good example; motivating other capable villagers to lead the changes; clear purpose for achieving common goals

(continued)

Table 2.1 (continued)

	A	B	C	D	E	F
Social networks & support for leadership	Local government and business networks, infrastructure (road, electricity, water, ICT, and financial services, know-how and professional advices)	Local government and business networks, infrastructure and other services, know-how and professional advice	Multi-level government support, business networks, financial support and infrastructure services, professional assistance	Obtaining professional advice from external professional institutions and experts, developing unique business networks, active engagement with government agencies and financial institutions	Accumulating network connections through the work outside village, actively engaging customers outside village, bringing expertise and support from professional and financial institutions	National and local government networks; international, national and local research institutes' networks, support of local infrastructure and financial services; support of village leaders and members
Leadership concerns about the well-being of followers	Improving household income, developing eco-friendly and sustainable agri-business, reserving *Bulang* minority culture for future generations	Reserving traditional culture, developing sustainable business for all villagers, looking after the well-being of women and children	Reserving and developing traditional culture and architecture, eco-development, increasing household income, developing beautiful & livable village	Achieving common prosperity, increasing village household income, maintaining forest and eco-system	Achieving common prosperity, making village life enjoyable and exciting, helping young people getting good education	Achieving common prosperity, making sustainable eco-system with a balanced approach of protecting forest and increasing household income

Leadership challenges for future business	Needs for improved infrastructure and services; competition from large-scale commercial production by investors from outside; attracting capable people to manage village-based business from outside; effective brand name and marketing strategies	Getting sufficient loans from financial institutions; difficult to get capable people to manage business from outside; difficult to build brand name	Loosely organised cooperative; barriers for marketing; lack of capable people from outside to manage business; traditional culture and village protection	Building brand name and reputation; sufficient infrastructure and financial services; attracting capable people to manage business from outside; protecting village environment and traditional architecture	Lack of sufficient financial support from formal financial institutions; attracting capable people to manage business from outside	Fully using carbon credit to benefit villagers, needs more technical support from professional institutions, continuing improvement of eco-diversity and sustainability
Leadership development	Needs more knowledge related to production know-how, marketing and tourism management	Needs more knowledge on marketing and developing brand name	Needs more knowledge on business management and marketing; eco-development and tourism management	Needs more knowledge on business management and marketing; eco-development and sustainable architecture and tourism	Needs more knowledge on business management and marketing; eco-development and sustainable architecture and tourism	Needs more professional people to support rural community development; needs more international and national level collaborations

fact, the case studies have demonstrated that social entrepreneurship is like any other entrepreneurship which involves risky business. In the early years, most of them had to take many risks in terms of getting a loan by using personal assets to secure a mortgage, finding a suitable business model and products, developing good networks and markets, dealing with multiple suppliers, clients and customers. Given the background that the rural economy was not based on a free market for many years, it was very hard for anyone to lead or drive change. However, these social entrepreneurs are very persistent when it comes to achieving their goals and are quick learners—they learn from negative experiences and look toward positive outcomes. All of the stories in regard to their processes of rural development reflect such a phenomenon; there is a traditional Chinese saying, that 'success is built based on past failure'. Therefore, we can see there is a clear learning curve amongst these social entrepreneurs to accumulate both negative and positive experiences along the way of social enterprises' development. In the latter stage of their development, they are more confident about future changes, knowing much more about what they should and should not do, and more capable of utilising different resources and social networks. The transformation can be seen in the example of a small project focusing on one or two areas of agri-business such as tea plantation and production, gradually expanding to more diverse business activities such as ecotourism and biodiversity of mountain products. At this later stage, the business operation requires more holistic strategic consideration and multi-skilling in terms of management and marketing. Therefore, improving knowledge and skills of social entrepreneurs themselves or recruiting capable people from outside become crucial for their future business success.

The third finding is about the support of internal and external stakeholders in terms of infrastructure development and other services (particularly knowledge and know-how, financial, and policy support), through social network building by the rural social entrepreneurs. Our case studies demonstrate that they are able to obtain support from villagers for land and forest reform and developing cooperatives at village level. In addition, they are also capable of obtaining support from external stakeholders, including the initial know-how, financial and policy support, infrastructure, services and marketing networks during the process of development. Key elements such as obtaining professional advice from external professional institutions and experts, developing unique business networks, actively engaging with government agencies and financial institutions have been defined by our cases as crucial for the success of rural cooperatives and communities' development.

The fourth finding is related to the process of social entrepreneurship from planning to developing and advancing rural cooperatives' development. This is a challenging and difficult process and requires the rural social entrepreneurs to be trained and developed in order to be more capable in terms of strategic and innovative thinking, to apply consistency and be persistent with their goals and plans, flexible to deal with changes, and lead the transformation with competence. Fundamentally, the social mission underpins their commitment to realise these goals and plans. Most of the cases identified that their major consideration is to enhance social well-being, this includes increasing villagers' income, developing eco-friendly and sustainable agri-business, preserving ethnic traditional cultures, and achieving common prosperity.

On the other hand, they are realistic about the current and future challenges and difficulties. On their journey they have learnt from negative experiences of failure. They have become more confident and capable of facing new challenges but aware of their limitations. Hence, many key areas have been identified for personal development and improvement, such as a need for more knowledge related to production know-how, management and marketing skills, capability to develop and preserve brand name, and even more international collaborations for rural communities and cooperatives' development. These qualities and understanding have given us confidence about the future development under their leadership and filled us with a sense of optimism and hope.

There are a number of meaningful implications for the conceptual understanding of social entrepreneurs and the development of rural communities. First, our findings provide the broader concept and definition of social entrepreneurship which emphasises innovative and exceptional leadership in social enterprises (i.e. village cooperatives), with the goal of achieving sustainable development and social mission (Weerawardena & Sullivan Mort, 2001) of common prosperity. Our sample village cooperatives and their leaders are good examples of profit-driven business working for community development and operating in a competitive economic market environment. Secondly, we have obtained sufficient evidence to demonstrate the social entrepreneurs' ability to attract and utilise different resources in a competitive environment and adopt innovative ways to create social value (Dees, 1998). Three key factors for successful social entrepreneurship, namely innovativeness, proactiveness, and risk-taking (Weerawardena & Sullivan Mort, 2006) could be found in our case studies. In fact, our cases show a combination of 'business skills' and 'social mission' is fundamental for a social entrepreneur to obtain and develop during the entire process (Sharir & Lerner, 2006). Finally, our analysis provides a meaningful outcome and enables us to create a framework regarding the leadership role of social entrepreneurs in relation to the competitive environment, the external and internal stakeholders, the support of infrastructure and other services as well as the way of managing the process of social enterprise development.

In addition, the practical implications are equally important. These social entrepreneurs could be vulnerable to market competition and sometimes the risk-taking could lead to personal sacrifice. They need both institutional and community support during the process, in terms of policy initiatives, resources, financial and even emotional support from other stakeholders. We also observed that the family support given to these social entrepreneurs during difficult times was crucial to their survival. In addition, providing continuous learning opportunities for these social entrepreneurs could build their competence to face the challenges, in particular related to knowledge of new products, technology and markets. More voluntary support from professional bodies and individuals in these areas would enhance these social entrepreneurs' capability and confidence.

Our research has shown some unique characteristics of social entrepreneurship in rural China. It could also have some meaningful implications for other countries, in particular in relation to developing countries in the transitional stage with mixed economic and ownership systems, limited financial and other resources, and potential market opportunities. Such environments require a more innovative and proactive approach towards the development of social enterprises. In this sense, social entrepreneurship plays a vital role in the process of community development as well as makes a significant contribution to nation building and common prosperity. This is an alternative business model from the mainstream of business operations under market capitalism which focusses on individual self-wealth building and expansion. It is hoped that, through our case studies, new ways of conducting business—based on village cooperative models—could lead to more equitable and mutually beneficial development paths whilst fostering stronger cooperation amongst citizens to achieve common goals for the entire society.

References

Dees, J. G. (1998). Enterprising nonprofits. *Harvard Business Review, 76*, 55–67.

Huang, Z. J., & Cui, P. (2009). Nongcun zhifu daitouren nengli jiegou ji xingcheng jizhi yanjiu [The research of rural leaders' capability structure and formation mechanism]. *Human Agricultural Sciences, 8*, 176–178.

Li, J., Young, M. N., & Tang, G. (2012). The development of entrepreneurship in Chinese Communities: An organizational symbiosis perspective. *Asia Pacific Journal of Management, 29*, 367–385.

Li, S., Schulze, W., & Li, Z. (2009). Plunging into the sea, again? A study of serial entrepreneurship in China. *Asia Pacific Journal of Management, 26*, 667–680.

Liu, B. (2004). Nongcun jingji fazhan yu nongcun zhifu daitouren [Rural economic development and rural leadership]. *Commercial Research, 292*, 180–181.

Sharir, M., & Lerner, M. (2006). Gauging the success of social ventures initiated by individual social entrepreneurs. *Journal of World Business, 41*, 6–20.

Wan, B. R. (2007). Yao zhongshi xinnongcun jianshe de liangzhi shenglijun [The importance of two forces for new countryside construction]. *Agricultural Economic Issues, 11*, 4–7.

Weerawardena, J., & Sullivan Mort, G. (2001). Learning, innovation and competitive advantage in not-for-profit aged care marketing: A conceptual model and research propositions. *Journal of Nonprofit & Public Sector Marketing, 9*(3), 53–73.

Weerawardena, J., & Sullivan Mort, G. (2006). Investigating social entrepreneurship: A multidimensional model. *Journal of World Business, 41*, 21–35.

Wu, J. H., Yu, H. L., & Huang, Z. J. (2008). Nongcun zhifu daitouren chengzhang de shihui ziben fengxi [The social capital analysis about the growth of the rural wealth-leader]. *Rural Economy, 24*(1), 95–97.

Yang, J. Y., & Li, J. (2008). The development of entrepreneurship in China. *Asia Pacific Journal of Management, 25*, 335–359.

Zeng, P. S., Li, D. X., & Yang, X. N. (2005). Guanyu qianfada diqu nongcun zhifu daitouren peiyu de sikao [The consideration of the training of rural leadership in less developed countryside]. *Chengren Jiaoyu* [Adult Education], *225*, 71–72.

Chapter 3
Market-Based Financing Mechanism and Design of Voluntary Trading Platform for China Forest Carbon Sinks

3.1 Introduction

Forests play a vital role in climate change as carbon sequestration through forestry-related mitigation options, and climate change can also affect the mitigation potential of the forest sector (i.e., native and planted forests) (IPCC, 2007). One of the great virtues of carbon storage and sequestration is that CO_2 is perfectly fungible and forests are one of the best 'sinks' for storing that carbon (Totten, 1999). There are abundant relatively low cost mitigation opportunities in forestry and therefore the recognition of these opportunities has generated proposals for forestry credits (Stewart, Kingsbury, & Rudyk, 2009).

According to the 2010 Global Forest Resources Assessment (FRA) published by the Food and Agriculture Organization of the United Nations (FAO, 2010a), the world's total forest area is over four billion hectares or 31 % of total land area, which corresponds to an average of 0.6 ha per capita. The FRA estimates that forests store 289 gigatonnes (Gt) of carbon in their biomass alone. Accordingly, a forest carbon market has developed rapidly as a step towards the stabilisation of greenhouse gas concentrations in the atmosphere to avoid global warming. Furthermore, not just developed countries but recently also developing countries have initiated their own domestic voluntary forestry carbon markets as part of the domestic portfolio of GHG emission reduction activities by actively developing market-based mechanism.

China, a developing country and the world's largest CO_2 emitter, promotes afforestation as a key strategy for climate change mitigation, and is increasing efforts to speed up the pace of plantations nationwide. In response to government calls to enhance forest carbon sink trading and develop the forest carbon market, in late 2011, Huadong Forest Exchange implemented the first forest carbon sink trading based on plantations in six different areas in China, all of them afforestation projects registered under the China Green Carbon Foundation (CGCF).

Y. Zhu et al., *Transforming Rural Communities in China and Beyond*,
DOI 10.1007/978-3-319-11319-7_3

However, like other developing countries, China has faced enormous deficits in financial resources for climate change actions within the forestry sector (FAO, 2012). Firstly, limited budgets and few financing channels are one of the key constraints for initiating afforestation projects and forestry related mitigation activities in China. Secondly, financing capacity through the global carbon market under the afforestation/reforestation Clean Development Mechanism (Afforestation/ Reforestation CDM) is limited, since there are only a few A/R CDM projects registered under the United Nations Framework Convention on Climate Change (UNFCCC). Thirdly, at present, the basic conditions are not yet ripe for the carbon market to finance forest carbon activities, as pricing and market mechanisms have not yet been completely established. Fourthly, the demand for forest carbon credits is significantly low and since the pilot transaction of the China Green Carbon Fund (CGCF), which belongs to a country-specific forest carbon trading scheme, there has been no consecutive exchanges reported.

As an essential component for funding forestry activities, China recognises the importance of encouraging the private sector to participate in the initiative of carbon funds in order to fill the gap of limited government budgets and to use market-based financing mechanisms. Therefore, it is imperative to study experiences and implementation of China's voluntary forest carbon market and its financing mechanism in order to enhance the capability of financing mechanisms and thus develop better programs.

The chapter highlights several key features of financing mechanism of Chinese domestic forest carbon programs and further analyses a trading platform for a pilot transaction of a Chinese forest carbon sink in the context of platform, supply side, demand side, pricing mechanism and transaction commodities with the focus on the CGCF. It hopes to extend experiences obtained in a pilot transaction of Chinese forest carbon sinks, for the benefit of other provinces of China, and explore possible implications for other developing countries.

3.2 Theoretical Basis of Forest Carbon Sinks, Forest Carbon Market and Domestic Forest Carbon Program

3.2.1 Previous Studies of the Financing Mechanism of Chinese Forest Carbon Sinks

As the voluntary forest carbon market is in the early stages, previous studies of the financing mechanism of voluntary forest carbon program in China have been published quite recently. From the perspective of forest carbon credit trading, Zhang and Ceng (2012) pointed out that, in the course of climate change negotiations, China's commitment to reduce emissions is to reduce carbon intensity; therefore, it will take time to form forestry carbon offset market under a cap and trade scheme. In this context, the current Chinese forestry carbon sequestration should focus on the promotion of the market and the establishment of voluntary market mechanisms, and its rules and service should adapt the needs of the voluntary market.

Hwang et al. (2011) stated that as the carbon market has to consider the carbon emissions rights the basis for trading at financial markets, a country without any cooperation with other countries, under the limited economies of scale, will find it difficult to develop carbon markets. South Korea, China and Japan have not only adjacent geographical advantages but also, from an economic point of view, there is the possibility of complement effects as well as cooperation within the climate change related financial sector.

Hwang et al. found inadequate financing in the Chinese voluntary forest carbon market and proposed multi-channel financing through drawing on traditional forestry financing methods, which are regarded most effective and also have great potential for financing. Based on expert interviews and structured surveys, Cao (2008) concluded that forest carbon markets represent a highly innovative, fluid or dynamic new financing mechanism for promoting sustainable forestry practices in Asia.

3.2.2 *Definitions of Concepts*

3.2.2.1 Forest Carbon Sinks

The concept of a 'sink' went through the following process of development during the course of negotiations at an international treaty, the United Nations Framework Convention on Climate Change (UNFCCC) since 1992. Although, the Kyoto Protocol[1] focuses mainly on emission reductions, the option to reduce the concentration of greenhouse gases in the atmosphere through the enhancement of terrestrial carbon sinks was debated ever since the Kyoto conference in 1997, where the subject entered into the negotiations as the Land Use, Land-use Change and Forestry (LULUCF) issue (Jung, 2003). Forest carbon sinks are the process, activity or mechanism for absorbing carbon dioxide from the atmosphere and combining it with carbon sink trading by means of afforestation, reforestation, forest management, reduction of deforestation and so on (Cao & Chen, 2010).

Among three mechanisms under the Kyoto Protocol, the only one relevant to developing countries was the clean development mechanism (Cao & Chen, 2010). Originally, sinks projects were not included in the CDM (Olsen, 2007). In Marrakesh 2001 it was decided that only afforestation and reforestation projects would be included under LULUCF activities. In the subsequent Bonn Agreement and Marrakesh Accords, the carbon sink project was considered as a CDM project, but limited to afforestation and reforestation (Cao & Chen, 2010).

In 2005, Reducing Emissions from Deforestation and Forest Degradation (REDD) was entered into the UN climate change discussions in Montreal. REDD was officially included in the 2007 Bali roadmap for further action and has since

[1] In 1995, countries launched negotiations to strengthen the global response to climate change and two years later adopted the Kyoto Protocol, which came into effect in February 2005 and offers three flexible mechanisms, namely Emissions Trading, Joint Implementation (JI) and CDM.

expanded into REDD+ which now includes Conservation and Enhancement of Forest Carbon Stocks, and Sustainable Management of Forest. Under the Bali Action Plan (BAP), both developed and developing countries will need to take nationally appropriate mitigation actions, known as NAMAs, to reduce their greenhouse emissions (Daviet, 2009). The expanded version serves the interest of developing countries that have been actively protecting and enhancing their forests.

3.2.3 Background on Global Forest Carbon Markets

3.2.3.1 Forest Carbon Projects in Regulated Markets Versus Voluntary Markets

Firstly, compliance carbon markets, which cover the bulk of transactions worldwide, are associated with national or international regulatory frameworks limiting greenhouse gas emissions, often imposing an upper limit (cap) on emissions. The compliance carbon markets include the Kyoto Protocol's CDM, Joint Implementation (JI), Kyoto Assigned Amount Units (AAU), the New South Wales Greenhouse Gas Reduction Scheme (NSW GGAS) and the New Zealand Emissions Trading Scheme (NZ ETS).

Secondly, voluntary carbon markets provide carbon benefits to businesses, individuals and households acting of their own volition (e.g. seeking to reduce their carbon footprint) (Valatin & Coull, 2008). The voluntary markets fall into voluntary Over-the-Counter (OTC) market and country-specific voluntary programs worldwide (Peters-Stanley, Hamilton, Marcello, & Sjardin, 2012). In addition to limited trading in compliance credits, they encompass all trading in carbon units issued outside statutory frameworks, which cannot be traded in compliance markets (Valatin, 2010).

3.2.3.2 Current Status of Global Forest Carbon Markets

Both trading volumes and transactions of forestry carbon credits have maintained growth, especially since 2008: the transactions showed a significant increase as exchanges; the CCX voluntary market and the New Zealand Emission Trading Scheme (Z ETS) registered its first forest carbon credits in 2007 and 2009 respectively; and the first Kyoto Protocol commitment period of 2008–2012 commenced. However, in 2011 there was a significant fall in volumes as offsets contracted over the counter in the international marketplace to voluntary buyers decreased. On the contrary, the market value was driven by domestic marketplaces like British Columbia's (BC) Carbon Neutral Government scheme and Australia's carbon price mechanism as well as the credits from A/R projects under the CDM, due to participants' willingness to contract credits ahead of the end of the Kyoto Protocol's first compliance period.

It is noteworthy that at present most forest mitigation activities take place in the voluntary carbon market. The 2012 voluntary forestry carbon credits accounted for 96 % (27 $MtCO_2$) of the total volume of forestry carbon transactions and 91 % (US$198 million) in value. This is because corporate buyers sought offsets from forestry to renew or pursue new climate targets and buyers in California and Australia also pursued offsets to prepare for compliance carbon markets (Table 3.1, Fig. 3.1).

Table 3.1 Volume and value of forestry carbon transactions

	Volume of transactions ($MtCO_2$)					Value of transactions (million dollar)				
Year	2008	2009	2010	2011	2012	2008	2009	2010	2011	2012
Total of voluntary forest carbon	5	–	27.5	18.3	27	36.8	–	126.9	185	198
Total of in-compliance forest carbon	0.2	–	2.5	7.4	1	0.3	–	6.6	52	18.1
Total of global forest carbon	5.2	21.5	30	25.7	28	37.1	87	133.5	237	216.1

Source: State of Forest Carbon Market (2009–2012), Ecosystem Marketplace, Hamilton et al. (2009), and Peters-Stanley et al. (2011)

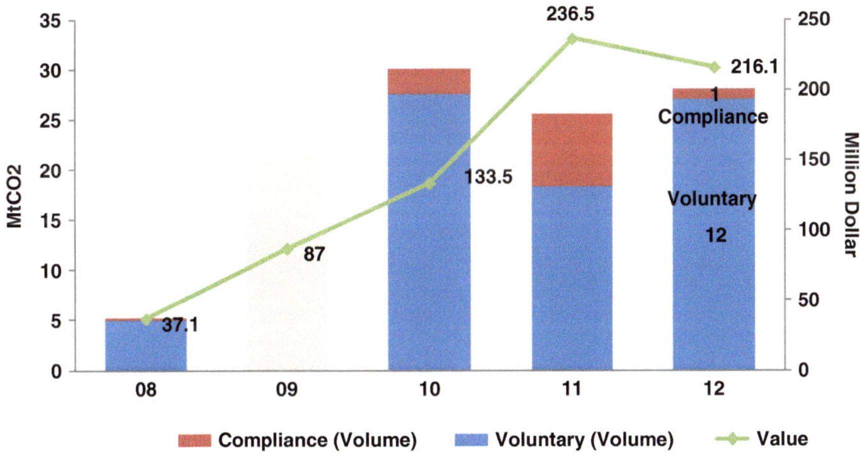

Fig. 3.1 Volume and value of forestry carbon transactions (Source: State of Forest Carbon Market (2009–2012), Ecosystem Marketplace)

3.3 A Voluntary Trading Platform for China Forest Carbon Sinks and Its Financing Mechanism

3.3.1 Background on China's Forest Carbon Sink Initiatives

China highly values the unique role of forestry in coping with climate change (UNFCCC, 2012). China's commitment on climate change and the policy on forest carbon sinks are backgrounds for forestry carbon transactions and also provide a favorable environment for the development of forestry carbon sinks.

3.3.1.1 China's Forestry Related Commitment on Climate Change

China's forest resources are still insufficient and there are issues of low quality and uneven distribution. Under this background, in order to mitigate climate change and to fulfill the obligations as a big country, China set afforestation as a key strategy. Though the forest coverage rate of China is 20.36 %, accounting for the world average of two thirds (the world's forest coverage rate of 30.3 %), per capita forest area of China is less than one fourth of the world per capita and per capita forest volume is only equivalent to 1/7[2] of the world's per capita stock volume (68.54 m^3).

In China, there has been a positive trend of forestry land area increase since the 1990s, benefiting from continuously strengthened forestry land protection and implementation of key forestry programs along with the in-depth promotion of national ecological rehabilitation policies (FAO, 2010b). For 20 years, from 1990 to 2010, although the rate of increase in forest area slowed, China's forest area has continued to increase - the diminishing rate of growth is 12.6 % (year 1990–1995), 9.1 % (year 1995–2000), 7.2 % (year 2000–2005) respectively. In addition, China has the world's fastest growing forest area and the largest national plantation (Table 3.2).

Table 3.2 Extent of forest and other wooded land

	Area (1,000 ha)			
FRA 2010 categories	1990	2000	2005	2010
Forest	157,140.6	177,000.5	193,043.9	206,860.6
Other wooded land	101,497.7	97,683.1	104,332.9	102,011.8
Other land	683,891.7	667,846.4	645,153.2	633,657.6
Of which with tree cover	n.a.	n.a.	n.a.	n.a.
Inland water bodies	17,470	17,470	17,470	17,470
Total for country area	960,000	960,000	960,000	960,000

Source: Global forest resources assessment 2010. Country Report, China

[2]China National Forestry Government Website, http://www.forestry.gov.cn/portal/main/s/201/content-495069.html

Although China remains a developing country, it alone accounted for 26 % of the world's total CO_2 emissions in 2010. China signed the Protocol to the United Nations Framework Convention on Climate Change (UNFCCC) on 29 May 1998 and ratified it 30 August 2002. China is not bound by the Kyoto Protocol and has not yet set a national cap on its greenhouse gas emissions voluntarily, however, due to the growing total GHG emission, China is facing and, in many cases accepting, greater pressure to act on climate change and thus intends to develop a carbon market.

In the context of China's 12th Five-Year Plan 2011–2015, Chinese leaders have set targets to further reduce energy intensity by 16 % by 2015. Along with measures to reduce pollution and increase the shares of non-fossil fuels in the energy sector, China has set goals to improve its CO_2 intensity by 40–45 % by 2020, with an interim target in the 12th Five-Year Plan of 17 % by 2015. Forestry dual growth objectives, namely to increase forest area and forest stock volume, is one of China's three objectives of voluntary reduction of greenhouse gases. By 2020 it is intended that forest area will be increased by 40 million hectares, and forest stock volume by 1.3 billion m^3 by 2020, compared with 2005 levels.

During the 12th Five-Year Plan period, China will continue to implement policies and actions on forest conservation and development. By strongly pursuing afforestation and sustainable forest management, the forest area will be enlarged by 12.5 million hectares and forest stock volume by 600 million m^3, striving to achieve the forest coverage of 21.66 %, to lay a solid foundation for meeting the targets for increasing forest area.

3.3.2 *China's Policy on Forest Carbon Sinks and Carbon Markets*

Title	Year	Contents
China Green Carbon Fund Management Interim Measures	Feb. 2005	It provides China's voluntary forestry carbon market with financing channels, management, monitoring and regulates funding of the CGCF and its use, management, recipient countries and monitoring by society and donors
Notice on pilot transactions on carbon emissions	Nov. 2011	Office of the National Development and Reform Commission (NDRC) issued and approved pilot transactions for five cities (Beijing, Tianjin, Shanghai, Chongqing, Shenzhen) and two provinces (Hebei and Guangdong) and initiated transactions in 2013
Voluntary greenhouse gas emissions transactions Interim Measures	June 2012	It carries out relevant provisions of China's voluntary carbon transactions and clearly states the six greenhouse gases to be included in the scope of transaction[3]

[3]http://www.hdlqjy.com/Archive/CarbonNewsDetail.aspx?id=2502

Currently, forest carbon trading regulations are based mainly on China Green Carbon Fund Management Interim Rule and Voluntary greenhouse gas emissions transactions Interim Measures, providing financing channels and related regulations for the non-Kyoto carbon market, which is a voluntary carbon offset market.

It is noteworthy that, in order to promote China's voluntary emissions trading activities in an orderly, standardised and effective manner, and gradually establish a cap-and-trade carbon emissions market, voluntary greenhouse gas emissions transactions 'interim measures' will provide policy support. The measures include general principles, management of voluntary emissions reduction projects, management of carbon emission reductions, transaction, validation and verification. After the implementation of the measures, carbon credits on the market will be approved by the National Development and Reform Commission (NDRC), thus market credibility and quality of the credits are likely to be increased compared to credits just verified by an audit agency.

China confirmed the establishment of its national carbon market's intentions in 2011, when the NDRC issued and approved pilot transactions for five cities (Beijing, Tianjin, Shanghai, Chongqing, Shenzhen) and two provinces (Hebei and Guangdong) and initiated transactions in 2013 (Peters-Stanley et al., 2012). China is actively promoting a domestic carbon trading pilot, while exploring the means of having afforestation and reforestation activities under the domestic carbon trading system[4] expanded to a nationwide carbon trading in 2015 (Hu & Tian, 2012). The voluntary forestry carbon platform is highly likely to be integrated into China's domestic cap-and-trade carbon market within 2020.

In summary, the Chinese Government is fully aware of the role of forests in mitigating climate change and promotes sustainable social and economic development with multiple effects and has taken strategic measures to develop forestry (UNFCCC, 2012) and the carbon market. All these indicate that the Chinese government has fully recognised the multiple effects of forests in climate change mitigation and adaptation, and in facilitating sustainable socio-economic development, and it has taken forestry development as a strategic option to cope with climate change (UNFCCC).

3.3.3 Key Features of China's Voluntary Forest Carbon Program

At present, forest carbon markets in China include three types of transactions: (1) forest carbon afforestation projects; (2) forest carbon afforestation projects initiated by the China Green Carbon Fund, in which certified credits are transacted

[4]http://www.jjly.gov.cn/page/colK_detail.php?wzid=5906&lbid=601

in voluntary carbon markets; and (3) A/R CDM projects under the Kyoto flexible mechanisms. As classified, the system of China's forest carbon market is different from other carbon emission trading schemes, as there is no cap-and-trade compliance regime and the demands of forest carbon sinks come voluntarily from previously stocked carbon credits and corporate or individuals buyers (Zhang & Ceng, 2012). Thus, this national domestic forest carbon trading system is self-imposed and separate from the Kyoto Protocol.

3.3.3.1 Funding Channel Is Limited and Still Not Mature

Firstly, limited budgets and few financing channels are one of the key constraints for initiating afforestation projects and forestry related mitigation activities in China. In particular, due to the absence of a compliance emission trading scheme, it is difficult to secure long-term market-based financing and carry out forest carbon sink service fully based on market mechanism. This could result in impediments to the development of forest carbon sinks (Huang, 2011).

3.3.3.2 Limited Financing Capacity Under A/R CDM

China is taking low-carbon matters into its own hands outside of the CDM. Forestry projects have faced a variety of hurdles under the CDM and this has resulted in China strategically choosing to develop domestic voluntary platform for forest carbon sinks. CDM has been accused of methodological complexities, high development costs, lack of up-front financing, taking too long to generate credits and small-scale marketplaces. This is apparent from the figures that China registered in its first A/R CDM project in 2006 and, up until recently, there are only five A/F CDM projects registered by the CDM Executive Board (EB) under the UNFCCC. There are 7,428 registered CDM projects, of which 52 forestry projects make up a meagre 0.7 % (as of 9 February 2014) (Table 3.3).

The European Union (EU) limits post-2012 eligibility of Certified Emissions Reductions (CERs) into the European Union Emissions Trading Scheme (EU ETS) to least-developed countries, a new rule that is scheduled to come into effect in the EU from 2013. These rules exclude the use, for compliance in the EU ETS (the biggest market for Kyoto credits) of CERs from projects other than those hosted in least developed countries (LDCs) or other countries with bilateral agreements with the EU, unless the projects were already registered in 2012.

Thirdly, the basic conditions are not yet ripe for carbon market for financing forest carbon activities, pricing and market mechanism advantages and its use is severely restricted (Guo & Nie, 2012) (Table 3.4).

Table 3.3 CDM forestry projects of China registered by the United Nations

Name of the projects	Facilitating reforestation for Guangxi watershed management in Pearl river basin	Afforestation and reforestation on degraded lands in Northwest Sichuan, China	Reforestation on degraded lands in Northwest Guangxi	Afforestation of degraded Shengie ecological zone in Helinge'er of inner Mongolia	Afforestation/reforestation on degraded lands in Southwest Sichuan
Project registered in CDM EB (UNFCCC)	10 Nov 2006	16 Nov 2009	15 Sep 2010	17 Jan 2013	5 Feb 2013
Types of projects	Afforestation/reforestation	Afforestation/reforestation	Afforestation/reforestation	Afforestation/reforestation	Afforestation/reforestation
Mode of project	Bilateral mode	Unilateral mode	Bilateral mode	Unilateral mode	Bilateral mode
Acres	4,000 ha	2,251.8 ha	8,671.3 ha	2,191.21 ha	4,196.8 ha
Amount of reductions per annum	25,795 MtCO2	23,030 MtCO2	87,308 MtCO2	6,725 MtCO2	40,214 MtCO2
Transaction price	US $4.35/t	US $5/t	–	–	–

Table 3.4 Comparison of A/R CDM and voluntary forest carbon sinks projects

Types	Afforestation/reforestation CDM in China	China voluntary forest carbon sink projects
Supply side	Cooperation between China and developed countries (three projects) and China as a sole investor (two project)	China
Demand side	UNFCCC Annex 1 countries, World Bank Bio Carbon Fund, Foreign enterprises	Domestic enterprises and individuals
Financing mechanism	World Bank Bio Carbon Fund, Equity investment, Loan	China Green Carbon Fund (Public carbon fund)
Mode of project	Bilateral mode[a] based (three projects) and two unilateral mode[b]	Unilateral mode
Type of project	Afforestation/Reforestation	Afforestation
Transaction price	US $4.5–5/t	US $3/t
Buyer motivation	Offsetting in-compliance carbon emissions, resale, investment	CSR, resale, investment

[a]It finds buyers in advance and two sides complete a project in cooperation
[b]Owners complete a project and then find buyers

3.4 Case Study: Pilot Exchange of Forest Carbon Sinks Under the Huadong Forestry Exchange (HFX)

3.4.1 Analysis on the China Green Carbon Foundation

Founded in July 2010, the China Green Carbon Foundation is the first nation-wide non-profit public funding foundation dedicated to combating climate change by increasing the number of carbon sinks in China. The precursor of this foundation is the China Green Carbon Fund (CGCF), which was established jointly by State Forestry Administration (SFA) of China, China Green Foundation, PetroChina, and Jiahan Forestry Investment Corporation in 2007. The functions of the foundation are to provide financial incentives to domestic actors helping to carry out afforestation projects, and to mobilise public and private entities for voluntary donations to public welfare activities for 'afforestation, increasing carbon sinks, participation in carbon compensation, and carbon footprint removal' among others (UNFCCC, 2012). In the past 3 years, the foundation has received several corporate donations totaling 300 million Yuan (around US$50 million), and planted more than one million mu carbon forests over ten provinces (region/city).[5] Moreover, the foundation became the observer of the UNFCCC Conference of the Parties in January 2013.

[5]http://baike.baidu.com/view/4231451.htm#refIndex_1_4231451

3.4.1.1 Properties of the Fund

The China Green Carbon Fund (CGCF), drawing on the international carbon fund, has the following purposes. First, the development of the CGCF is followed by two stages: public and commercial projects. When afforestation projects are going well and generating carbon sinks via plantations, donors will be given certified carbon credits prorate their financial participation in the fund. Unlike credits created from commercial projects, the credits generated from public projects are solely for voluntary purposes such as carbon footprint, carbon offsets and social corporate responsibilities, thus not allowed to be traded in the exchange. At the commercialisation stage, a project developer sells the carbon credits from a project to a prospective buyer. At this stage, the fund enters the carbon market and a professional fund management organisation. The ownership of credits purchased through the exchange belong to buyers, thus the secondary exchange can be completed.

Second, a long-term and effective mechanism of carbon forestry investment has not yet been formed. When the conditions are ripe, after the completion of commercial transactions, ownership of carbon sinks goes to the buyer, so the secondary transaction can be conducted. The revenues generated by the trading is returned to the fund and subsequently used for implementing forest-based carbon sink projects.[6] As both public and commercial projects are completed, an initial phase of voluntary forest carbon market is formed. The third stage, the exit mechanism of the fund, has not yet been established.

3.4.1.2 Sources of the Fund

Till late 2012, the CGCF has received US$3.03 billion (0.5 billion Yuan) (based on 1 Yuan equivalent of US 6.06 dollar as of 9 February 2014) from 1,500 businesses and 20,000 individuals, which creates approximately 120 mu acres of afforestation project based on plantations in 20 provinces. The sources of the fund are[7]:

- Individuals, corporates or donation by other organisations and communities:
- Domestic enterprises, organisations and individuals make donations to the Fund. The foundation received US$2.42 billion from local enterprises and four-fifths of the donation was contributed by state-owned companies combining 0.3 billion Yuan by the China National Petroleum Corporation (CNPC), a national gas corporation, and 0.1 billion Yuan by China Guodian Corporation (UNFCCC, 2012), the largest integrated energy company. Among the fund donated by the CNPC, approximately US$1.21 billion (67 %) was spent on energy forestry plantation projects and the rest, 33 % of the fund, was used in afforestation projects. In addition, up until 2011, the foundation has collected about four million RMB from individuals;

[6] http://www.forestcarbonportal.com/project/china-green-carbon-fund

[7] http://www.thjj.org/about-rule.html

- Income by organising communal activities of donation;
- Sponsorship by the government;
- Revenue by investment and increment of interest.[8]

3.4.1.3 Incentive Mechanism

Corporations donate to the China Green Carbon Fund afforestation, plantations are owned by local farmers, while enterprises obtain carbon sinks. As discussed in Chap. 4, tree plantings are undertaken on individually-managed lands or community lands, which are under the management of natural villages, the subunits of administrative villages in China. Therefore, the individual farmer households or natural villages claim the rights of timber products, while the donators can claim rights over carbon credits. Corporate donations to the Green Carbon Fund enjoy direct income tax free concessions. According to the People's Republic of China (PRC) Enterprise Income Tax Law, Article IV, (2), there is 25 % reduction in income taxes of the donated money; Article IX stated: 'in case of charitable donation occurred, the amount less than 12 % of the total annual profit may be deducted in the calculation of taxable income'.

3.4.2 Overview and Case Site Selection

In Zhejiang province in late 2011, Huadong Forestry Exchange (HFX) and CGCF implemented the first forest carbon sink trading based on plantations in six different areas in China, all of which were afforestation projects registered under the CGCF. The project activities were first initiated in 2008, 3 years before its first pilot trading for forest carbon sinks. Over 50 Chinese enterprises, ranging from large enterprises, forestry companies to investment companies, were involved in voluntary trading. At the launch of the pilot forestry carbon sink program in Yiwu city, Zhejiang, 11 enterprises and one individual finally registered to offset 148,000 t of carbon sink. Since then, the Chinese voluntary forestry carbon sinks transaction model has become more streamlined. However, since the first pilot transaction in 2011, there has been no consecutive forestry carbon sinks transactions reported and the Chinese forestry carbon market has entered a short period of stagnation.

[8]China Green Carbon Fund can be deposited in the interests that financial institutions charge and can also be an effective way to purchase securities, etc., to promote their safety and increase the value. The value-added incomes are still included in the fund in order to carry out activities related to the project.

3.4.2.1 Emissions Trading

Though the plan is in the early stages, Hangzhou will possibly be China's eighth region to launch an ETS. The 'Hangzhou energy consumption process equivalent carbon emissions trading Interim Measures (draft)' (hereinafter referred to as the 'Rules') proposed carbon emission reductions through 'carbon emission transactions'. This means that Hangzhou will try to replace the executive order with market behaviour and control carbon emissions regulation. In the future, operating enterprises are required to purchase annual carbon credits from the Government. If the targets are exceeded, the administrative department in charge of energy conservation will take measures such as limiting and stopping production of the enterprises. Enterprises will be given a cap and large emitters will need to purchase carbon emission rights to meet the goals of annual emissions. If the local government moves ahead with its plans, carbon-emitting facilities such as power stations and factories would be given caps on their CO_2 emissions and forced to buy emission permits to match annual emissions.

3.4.3 Analysis

The following five principles for ETS will now be discussed: (1) platform; (2) supply side; (3) demand side; (4) pricing mechanism and transaction methods; and (5) transaction commodities.

3.4.3.1 Trading Platform

The aim of the domestic voluntary platform for forest carbon sinks is to provide buyers (donators) with timely liquidity or previously stocked carbon credits, and a platform for achieving GHG emission reduction targets. Furthermore, it aims to provide quantitative compensations through market-based mechanisms and enhance forestry development with the support of the carbon finance market as a financing mechanism (Zhang & Ceng, 2012).

Located in the heart of Hangzhou City, the HFX is such a trading centre for forest based carbon credits. HFX was formed in November 2010 and has since become the leader on forest right system reform in Zhejiang Province, having established an integrated, standard and public forest resource trading platform over the region. Its service roles include forest right trading, forest products trading, forest resource asset appraisal, forest financial products innovation, forest carbon trading and so on. It has developed an integrated forest right trading system, consisting of trading rules, trading certificates, trading platform, information disclosure, and supervision mechanism. Meanwhile, tangible and intangible forestry markets construction has

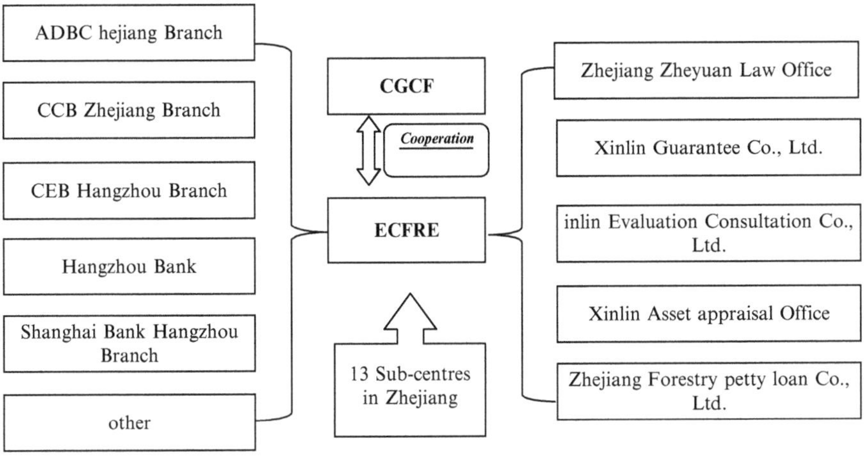

Fig. 3.2 Institution-building and cooperation (Summary based on http://www.hdlqjy.com/)

sped up thanks to information-based measures. HFX will be a regional exchange not only providing services to Zhejiang but also to the surrounding provinces.[9]

The development of forestry carbon sequestration in Zhejiang province has received considerable support from financial institutions. The publicity of the forest carbon sink trading platform is the critical factor for Zhejiang to further develop the forest carbon sink markets as it is initiated by the government, primarily financed by public funds. In order to ease the difficulty in financing medium-sized and small enterprises, Zhejiang Xinlin Guarantee Co. Ltd. (short form 'Xinlin Guarantee') was founded in August 2004. This is the first policy financing guarantee organisation in the forest industry in the whole country, while HFX is the leading organisation attracting investment by Xinlin. In addition, China Construction Bank (CCB) Zhejiang Branch, Shanghai Bank, China Everbright Bank (CEB) and other financial institutions, as well as the financial office of the Zhejiang Government, HFX and so on, have participated in forestry development in Zhejiang (the left side of Fig. 3.2 shows the financial institutions collaborating with the Forestry Exchange (FX)). Taking CCB Zhejiang Branch for example, it has designed credit products directed at forestry enterprises, including Forestry Rights Mortgage Loans and other green financing. Figure 3.2 shows institution building and cooperation in HFX, while Fig. 3.3 shows the forestry financing process.

3.4.3.2 Supply Side

The China forest carbon sinks are based on six different afforestation projects led by the CGCF. These are outlined in Table 3.5 (Table 3.6).

---

[9] http://www.hdlqjy.com/Archive/OtherInfoListDetail.aspx?id=204

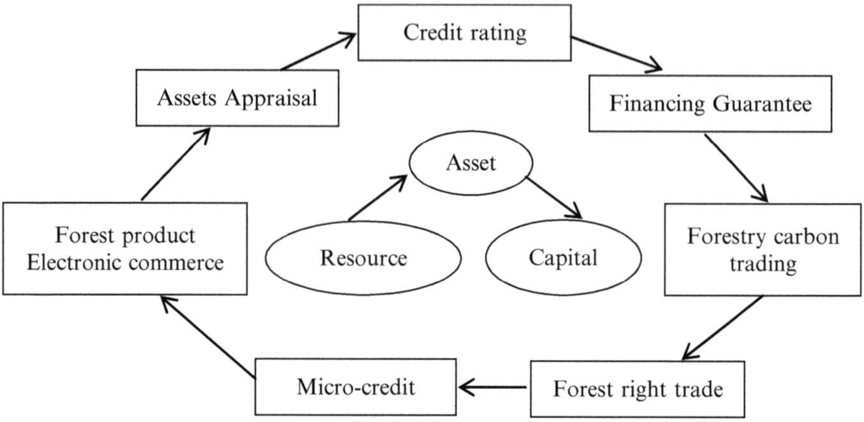

Fig. 3.3 Forestry financing process (Zhejiang Daily (2012-12-16) the 8th page Xinlin Guarantee: Finance gives service to forestry promoting green increase)

The crediting period is 20 years, from 1 January 2008 to 31 December 2027, which is the total number of years that the project may generate forest carbon sinks. In total, 148,571 t of carbon dioxide equivalent was transacted at the price of US$3/t CO2e (RMB 18 Yuan) and the price per unit was capped at US $6/t CO2e (RMB 36 Yuan). The initial forest carbon sinks were priced though a negotiation between buyers and sellers. The price is relatively low in order to promote forestry carbon sequestration transactions given that the cost of measuring and monitoring is approximately 30 RMB even though there are differences among each region and project.

3.4.3.3 Demand Side

The prosperity of the carbon market depends largely on demand but the lack of demand has impeded the development of China's forest carbon market. There are multiple motivations for buyers purchasing forest carbon sinks. Buyer demand comes from domestic corporates seeking to offset emissions and promote sustainable development for corporate social responsibility purposes and resale reasons. At the current stage of the voluntary carbon trading market, although the social values of buyers' motivations out-weight economic values, it is noteworthy that resale purposes are also one of key factors for private sector buyers purchasing forest carbon sinks. This is because although Chinese forestry carbon sinks can now be bought at a relatively low market price, prices may increase as more stringent climate change policy targets are introduced. This is also consistent with the global forest carbon market trend that companies purchasing credits for resale (typically offset retailers, wholesalers and other intermediaries) were the largest source of

Table 3.5 China forest carbon credit afforestation projects initiated under the China Green Carbon Fund, 2011

Name of Projects	Guangdong Heyuan Longchuan FCCAP	Guangdong Shantou Chaoyang Sihu FCCAP	Zhejiang Lin'an moso bamboo FCCP	Beijing Fangshan FCCAP	Gansu Dingxi Anding FCCAP	Gansu Qingyang State-owned Heshui Farm FCCAP
Area (ha)	200	200	46.67	133	133	133
Total investment	US$250,000	US$250,000	–	US$167,000	–	US$167,000
Volume of carbon credits generated	57,254 t of carbon dioxide equivalent	60,610 t of carbon dioxide equivalent	8,155 t of carbon dioxide equivalent	6,495 t of carbon dioxide equivalent	4,300 t of carbon dioxide equivalent	11,757 tonnes of carbon dioxide equivalent
Afforestation cost per one tonne of credits	US$4.4/t	US$4.2/t	–	US$25.7/t	–	US$14.2/t

FCCAP: Forest Carbon Credit Afforestation Project

Source: HFX, website of forestry divisions

Table 3.6 China forest carbon sink supplier, 2011

Types of project	Afforestation
Fund	CGCF
Types of carbon unit	Verified
Project start date	2008
Acres of afforestation	845.67 ha
Profit of credit transaction per ha	\$527/ha
Transaction date	2011
Volume supplied	148,571 t
Cost of accounting and monitoring	Approximately \$5/t on average
Transaction price	\$3/t
Value	US\$0.445 million
Volume purchased domestically	14,8571 t
Crediting period	20 years (2008–2027)

Source: HFX

private sector demand for credits contracted from developers in the primary market. The design of a future mandatory cap-and-trade scheme might allow initial buyers to resale credits at a higher price to entities regulated under the scheme. In fact, the pilot transaction of forest carbon sinks was made at the price of US3\$/t CO2e, which is relatively a low price compared to US\$4.35/t CO2e for CDM reforestation for Guangxi watershed and other A/R CDM projects registered to the UNFCCC.

From the industry perspective, buyers currently purchasing forestry carbon sinks include forestry, construction, manufacturing and financial services, motivated by Corporate Social Responsibility (CSR) and resale value. This indicates that unlike other mature forest carbon markets, high-emission industries, such as national oil companies, steel, chemicals and utilities, have yet to join the forestry carbon market.

3.4.3.4 Transaction Method and Pricing Mechanism

The pricing mechanism has not been established. The exchange method of the first pilot transaction of CFCS is an agreement for transfer, the transaction price being 18 RMB per tonne. In addition, in order to restrict price bubbles and abnormal transactions, the price is controlled under 38 RMB. Pricing was significantly lower than the cost price of measuring, primarily because it is generated by the two sides after the auction within the price range that the buyer and seller all can accept, thus buyers are likely to be satisfied with the prices.

The current pricing method is not conducive to sellers, since the buyer is not willing to bear the cost of plantations and other benefits, and only purchases the carbon credits. Furthermore, the payment method is not conducive to the seller. As shown in Table 3.7, buyers pay through installments, the first installment being 60 % of the total, which negatively affects the profits of sellers and delays the

Table 3.7 Transaction and payment of forest carbon sinks

Transaction method	Agreement for transfer
Transferor	China Green Carbon Fund
Transferee	11 enterprises and one individual
Payment method	Payable by installments, first installment is 60 % of the whole installments

Source: HFX

carbon revenues of investors. In addition, the payment industry is not conducive to the seller. For example, the practice of paying in installments, with the first installment only 60 %, is not conducive to the seller's recovery gains, and the investor's carbon revenue is delayed.

In short, the existing pricing method seriously deviates from the law of market value, and the payment of forestry carbon sinks projects also makes it difficult to obtain funds needed for implementation of the project through carbon trading; market-oriented financing mechanisms have not been implemented.

3.4.3.5 Transaction Commodity

The transaction commodity is divided into forest carbon sinks in the primary market (project developers' initial contracts) and forest carbon sinks derivatives in the secondary market. At present, China's voluntary forest carbon trading platform can only serve at the project-based transactions. Design of secondary market and trading product development and development and innovation of transaction commodities are necessary.

3.5 Challenges and Strategies for Market-Based Financing Mechanism

3.5.1 Challenges for Forestry Carbon Sink Trading

In China, as well as around the global, the development of domestic voluntary forest carbon markets and trading emissions for forestry sinks is still in the early stage. Currently, the following challenges need to be addressed for continuing program development of forest carbon sink projects with the support of financing mechanisms:

- ***Uncertainty about domestic demand due to unclear future national guideline***: There is an uncertainty about domestic demand due to unclear future national guidelines. The experience of New Zealand implementing its domestic voluntary carbon trading program suggests that, under the domestic voluntary platform, demand could be complicated by a set of national guidelines for offsetting and carbon neutrality claims that are laid out by the relevant departments. It also indicates that the primary future demand coming from domestic corporates will be highly influenced by the stringency of national regulatory frameworks and the evolution of international frameworks (Gong & Li, 2006). A highly domestic market might hinder international buyers' participation. Given the increasingly localised nature of offset supply and demand, developers on the outside would find it difficult to tap into domestic-only market opportunities. This is because carbon credits registered to domestic standards are not likely to be accepted in international carbon markets. Divergent rules, methodologies, and registries created under each domestic platform would increase the global transaction costs of offsets, which possibly expose international project developers and buyers to a certain degree of risks.

- ***Unclear market prospects for voluntary forest carbon market under country-specific trading platform***: In the absence of relevant international agreements, countries are developing independent and national-level mechanisms and it is unclear how these will fit under any future international framework. Furthermore, from the perspective of the domestic carbon market, it is also unclear how to monetise forest carbon offsets within the evolving GHG emissions trading schemes in China.

3.5.2 Strategies of Market-Based Financing Mechanism of China's Voluntary Forestry Carbon Trading

3.5.2.1 Establishment of Effective and Sustainable Forestry Carbon Fund

Establishment of financing mechanisms should offer investors appropriate enticements to take the risk of providing early forest carbon project funding and exit mechanisms, to guarantee the desired returns of private investors within expected timeframes. In order to develop the forestry carbon fund, the authors make the following suggestions.

First, incentive systems for companies participating in donation to carbon funds should be enhanced.

At present China Green Carbon Fund only provides tax incentives for the companies contributing to the fund, but as the forestry carbon sinks market launches in other provinces, there is an increasing need for a carbon fund. Design and use of a variety of incentives to encourage corporates to donate is an important strategy of forest carbon fund.

Second, private equity funds for upfront forest carbon plantation projects should be set up. Over the past 10 years, the standard of forestry investment funds has generally achieved more than 14 % annual return, significantly higher than S & P 500 index during the same time period, and the performance is slightly lower than the stock market volatility but offers stable long-term benefits (Totten, 1999). Although the international carbon funds are now quite mature, China has not yet established carbon funds, which are financed through multiple channels by institutions, enterprises and governments and makes consecutive transactions available; it remains a donation-based fund.

Third, China should establish exit mechanisms for Chinese carbon investment funds. Voluntary donors such as climate investment multilateral partners have begun to actively explore steering donated public funds to market mechanisms in order to attract more private financing flows and as a main exit mechanism of the voluntary donation for forest carbon activities. However, China's carbon funds are in its early stage, so that international carbon funds investing in projects in China mostly look for exit channels in foreign countries and gain investment income and profits (Ma & Chen, 2011).

3.5.2.2 Establishment of an Effective and Flexible Market-Based Financing Mechanism

The success of a market-based financing mechanism depends on the development of a mechanism of supply/demand and pricing, etc. Thus, from the perspective of the design on trading system, multiple detailed strategies can be drawn as follows in order to establish both effective and flexible market-based financing mechanism.

Enabling Forestry Carbon Credits to Offset Enterprises' Emission Reduction Targets

A key reason why Chinese voluntary forestry carbon sinks transactions are currently facing a stagnation period is that China has no obligation to establish a compliance carbon market, and companies are not obligated to purchase carbon emission rights, neither of which are offset by forestry carbon credits in the market. Therefore, most companies lack motivation to buy forestry carbon credits.

A strategy to encourage Chinese large GHG emitting enterprises to take part in the transaction of carbon credits is one of the most effective ways to induce private sector investment and enlarge possible market demands. The buyers of pilot forest carbon credits do not yet include large GHG emitting enterprises, for example, enterprises in electric power, iron and steel, cement, plastic, aluminum smelting and mechanical processing industries. When China's enterprises are given a cap, it will generate demands of enterprises purchasing carbon credits to offset carbon emissions. High energy-consuming enterprises will be the main buyers of the future forest carbon sinks, which will be the inevitable future trend.

Introducing Forward Emission Reductions Purchase Agreements

Voluntary forest carbon credits can be sold both to spot and forward contracts. However, at present, there are only spot contracts available in China's voluntary forest carbon trading platform, where certified credits are usually paid for upon delivery.

'Forward' refers to the fact that parties agree to the purchase and sale of Verified Emissions Reductions (VERs) that have not yet been created (Hawkins et al., 2010).

As far as China's voluntary forest carbon trading is concerned, due to the long growth cycle of trees, the generation and transfer of forest carbon sinks also needs to go through a long cycle, inducing supply sides to bear a great amount of investment pressures and risks. Thus, forward purchase agreements provide security for the supplier in terms of up-front financing for forest carbon project activities and can significantly reduce the project risks to investors or suppliers. Second, a forward purchase agreement can be relatively standardised, yet can be customised for specific transactions and well-suited for use in a guidance document.

Third, forward purchase agreements can include forest carbon specific issues related to project activities, advance payments, default and remedies clauses, and other issues, in way in which spot purchase agreements cannot.

However, as China's voluntary forest carbon trading is in its early stage, the introduction of forward sales should account for fundamental uncertainties in timing and the amount of carbon credits that will be generated by a forest carbon project (Wu et al., 2009). Therefore, initially these types of contracts are likely to be available only in relatively mature forest carbon markets. In short, it provides a contract framework, enabling buyers and sellers to discuss the issues and negotiate contracts, which to some degree are unique or specific issues of forestry carbon sinks.

Linking Emissions Trading Schemes

The future global carbon market is highly likely to be combined. China therefore can attempt to establish mutual agreements through a common crediting mechanism such as forest carbon sinks with other international markets. Also, China can attempt to establish bilateral green carbon funds with other Asian Pacific countries and develop bilaterally agreed methodologies.

The forest carbon platform will be integrated into an emerging national GHG regime of China and gradually attempt to link them to the international rules and markets. Given the current development stage of China's voluntary forest carbon trading platform, direct bilateral linkages between China and EU ETS will be difficult. However, China is likely to explore new mutually approved methodologies in cooperation with other countries, such as monitoring, reporting, and verification of forest carbon sinks, banking, registries, methodologies, transaction rules, etc.

Secondly, as carbon markets are characterised by carbon-traded financial markets, one country, without any carbon partnership with other neighboring countries and given the limited economies of scale, will find it difficult to expand market size and liquidity. The size and demands of China's domestic carbon market is not sufficient to establish an international carbon exchange. South Korea, China and Japan have not only adjacent geographical advantages, but also from an economic point of structure, there is a possibility of complement effects as well as cooperation on the climate change related financial sector.

Thirdly, in the short-term, indirect linking schemes are the most prominent (Yang & Chen, 2011). China is able to cooperate with other countries in terms of carbon trading and carbon finance through forestry carbon sink transactions as a starting point. The feasible process for linking trading schemes internationally and its prospect effects can occur in two phases. The first phase of globalisation of the carbon market is that China will establish a GHG emission scheme prior to 2020 and integrate carbon offset mechanisms such as forest carbon sinks under cap-and-trade carbon market. Therefore, China should attempt to establish bilateral green carbon funds with other Asian Pacific countries in order to implement plantation projects and develop bilaterally agreed methodologies. A successful linking scheme should meet a list of conditions, such as that bilateral or multilateral countries should have reached mature state and effectiveness, and have close economic ties and history of policy coordination (Stewart et al., 2009). For example, China and South Korea signed the MOU on cooperation for North East Asia carbon partnership in 2009, however, there has been no further discussion or practical action plans for the carbon partnership. Meanwhile, the Korea Exchange will host South Korea's greenhouse gas emissions trading scheme, which is due to be launched on 1 January 2015. Within this scheme, more than 400 of South Korea's largest polluters will face their emissions goals. Under the compliance market, forest carbon sinks offsetting carbon emissions can be an effective way of meeting emission reductions goals for those Korean enterprises. Thus, the larger the Asia Pacific offsetting market that can be created, the more efficient its reduction efforts and cost savings will be made.

Subsequently, at the second phase of globalisation of the carbon trading market, China will be likely to link other national trading schemes. Thus, China's carbon trading market will be developed as an international carbon financial center, not only making the domestic carbon trading market seize the initiative in the international carbon trading market pricing, but also greatly enhancing bargaining power of South Korea, Japan and other Asia Pacific countries in the international climate change negotiations.

Conclusions

Since the establishment in 1992 of the UNFCCC, forestry has been the critical subject of climate change negotiation. Thus, numerous local and national governments initiated voluntary offset projects and started participated in trading of forest carbon sinks. Furthermore, currently forest mitigation activities take place in the voluntary carbon market, while the rise of country-specific programs is a significant new trend in the international carbon market.

(continued)

In this context, China set forestry projects and its trading market as a key strategy to mitigate climate change and is vigorously promoting the domestic carbon trading pilot. In 2007, the foundation of the China Green Carbon Fund opened a platform for Chinese corporates, organisations and individuals to take part in transaction of forest carbon sinks. Subsequently in 2011, the first forest carbon sinks generated from six different afforestation projects was transacted in the HFX. The domestic scheme is a voluntary market and different from A/R CDM under the Kyoto Protocol that the supply and demand of forest carbon sinks are entirely domestic based.

However, China is facing challenges of limited budgets and few financing channels. In addition to this, the voluntary forest carbon market of China is in its early stage as China's voluntary forest carbon trading platform can only serve at the project-based transactions. This indicates that in order to promote and develop voluntary forest carbon market, more resources will be needed, in particular, a flexible and market-based financing mechanism and its design.

Thus, multiple strategies for market-based financing mechanisms can be drawn in two aspects—forest carbon funds and market-based financing mechanisms. As far as forest carbon funds is concerned, design and use of a variety of incentives for corporates and any other participant is important as well as providing early forest carbon project funding and exit mechanisms to guarantee profits of investors. From the perspective of the market-based financing mechanism, a scheme that enables forest carbon sinks for corporates to offset emission reduction targets, an introduction of forward emission reductions purchase agreements and linking emissions trading schemes should be taken into account.

In summary, a market-based financing mechanism and its design has become a critical issue for China's forest carbon market. Thus, China is highly likely to develop China specific market-based financing mechanism and design it by combining domestic and international carbon markets and trends. Financing mechanisms for the forest carbon market should be highly innovative and flexible, which secures sustainable program development of forest carbon sink projects and enables China to enhance its stance at international climate negotiations.

This chapter shows several methods to promote a sustainable forestry financing mechanism in China as well as possible implications for other developing countries preparing for or currently developing trading platforms for forest carbon sinks at a national level.

References[10]

Cao, K. (2008). Thoughts on financing mechanism of China forestry carbon market. *China Business Magazine*, No. 2.

Cao, M., & Chen, Y. (2010). Forest carbon sinks and biodiversity conservation from China's perspective. *Reviista Catalama De Dret Ambiental, 1*(2), 1–21.

Daviet, F. (2009). *Beyond carbon financing: The role of sustainable development policies and measures in REDD*. Washington, DC: World Resources Institute.

Food and Agriculture Organization of the United Nations (FAO). (2010a). *Global forest resources assessment* (FAO Forestry Paper 163). http://www.fao.org/docrep/013/i1757e/i1757e00.htm. Accessed 12 Nov 2013.

Food and Agriculture Organization of the United Nations (FAO). (2010b). *Global forest resources assessment: Country report China*. http://www.fao.org/forestry/20302-0574060b798cf421dcbca8b5aa77a2448.pdf. Accessed 15 Nov 2013.

Food and Agriculture Organization of the United Nations (FAO). (2012). *Forest management and climate change: Stakeholder perceptions* (Forests and Climate Change Working Paper 11). Rome. http://www.fao.org/forestry/climatechange/en/. Accessed 10 Oct 2013.

Gong, Y., & Li, N. (2006). Analysis on demand and design of China forestry carbon sinks projects. *Forestry Economics*, No. 6.

Guo, S., & Nie, Y. (2012). Financing of forestry carbon sinks; the environment, capacity and strategy. *Forestry Economics*, No. 12.

Hamilton, K., Sjardin, M., Shapiro, A., & Marcell, T. (2009). *Taking root and branching out*. Ecosystem Marketplace. http://www.forest-trends.org/documents/files/doc_2384.pdf

Hawkins, S., Nowlin, M., Riberio, D., Stoa, R., Longest, R., & Slazman, J. (2010). *Contracting for forest carbon: Elements of a model forest carbon purchase agreement*. Durham, NC: Duke Law School.

Huang, Z. (2011). *Study on financing mechanism of voluntary forest carbon market*, No. 12. Hangzhou, China: Forestry Division of Zhejiang Province.

Hu, Y., & Tian, Z. (2012). Development paradigm of China forestry carbon sink and its solutions. *Quest*, No. 11.

Hwang, Y.-S., Choi, Y.-J., & Lee, Y. (2011). International trade: Collaboration for carbon market of three countries: KOREA, JAPAN and CHINA. *International Region Research, 15*(2), 427–447 (in Korean).

IPCC. (2007). *Fourth assessment report: Climate change 2007*. http://www.ipcc.ch/publications_and_data/publications_ipcc_fourth_assessment_report_wg2_report_impacts_adaptation_and_vulnerability.htm. Accessed 12 Nov 2013.

Jung, M. (2003). *The role of forestry sinks in the CDM: Analyzing the effects of policy decisions on the carbon market* (HWWA Discussion Paper, No. 241). Hamburg, Germany: Hamburg Institute of International Economics.

Ma, Y., & Chen, B. (2011). *State of China carbon fund and operating mechanism* (Serial No. 211–10, pp. 131–135). Academic Exchange, http://www.cnki.com.cn/Article/CJFDTotal-XSJL201110030.htm. Accessed 15 Nov 2013.

Olsen, K. H. (2007). The clean development mechanism's contribution to sustainable development: A review of the literature. *Climatic Change, 84*, 59–73.

Peters-Stanley, M., Hamilton, K., Marcello, T., & Sjardin, M. (2011). *Back to the future: State of the voluntary carbon markets 2011*. Ecosystem Marketplace. http://www.forest-trends.org/publication_details.php?publicationID=2828

Peters-Stanley, M., Hamilton, K., & Yin, D. (2012), *State of the forest carbon markets 2012*. Ecosystem Marketplace.

[10]Note: Newspaper and magazine articles quoted in the footnotes are not included here.

Stewart, R. B., Kingsbury, B., & Rudyk, B. (2009). *Climate finance for limiting emissions and promoting green development: Mechanisms, regulation and governance*. Public law & legal theory research paper series (Working Paper No. 09–66). New York University Abu Dhabi Institute.

Totten, M. (1999). *Getting it right: Emerging markets for storing carbon in forests*. Washington, DC: World Resources Institute.

UNFCCC. (2012). *Second national communication on climate change of the People's Republic of China*. http://unfccc.int/resource/docs/natc/chnnc2e.pdf

Valatin, G. (2010). *Forests and carbon: Valuation, discounting and risk management*. Research Report, http://www.forestry.gov.uk/pdf/FCRP012.pdf/$FILE/FCRP012.pdf

Valatin, G., & Coull, J. (2008). *Payments for ecosystem services: Findings and perceptions from the USA*. Edinburgh, Scotland: Forestry Commission, Forest Research.

Wu, S., Zhang, X., & Song, W. (2009). Compensation standards of global voluntary forest carbon market, *Forestry Science*, No. 3.

Yang, Z., & Chen, B. (2011). Carbon market trend and globalization strategy of the EU carbon finance. *Economic Review*, No. 1.

Zhang, X., & Ceng, Y. (2012). Establishment of China forestry carbon sinks management mechanism. Forestry Economics, No. 8.

Chapter 4
Carbon Trade, Forestry Land Rights, and Farmers' Livelihood in Rural Communities in China

4.1 Introduction

Under the government promotion, development of forest carbon sequestration projects has made great progress in China. These mainly include Clean Development Mechanism (CDM) projects under the framework of the Kyoto Protocol and forestry carbon sequestration in a voluntary market, such as the China Guangxi Pearl River Basin Reforestation Management Project approved by the United Nations CDM Executive Director in November, 2006 (UNFCCC, 2006). However, CDM forest carbon sequestration projects under the Kyoto Protocol have very demanding requirements on woodland eligibility, in addition to those for additionality and baseline, so there are rare projects that can conform to its standards and be suitable for development (Thomas, Dargusch, Harrison, & Herbohn, 2010). Consequently, the policy has been unable to meet its objectives of significant carbon emissions reduction, development of ecological forest and increasing farmers' income. Therefore, the Chinese government began to give its vigorous support to forestry carbon sequestration in a voluntary market.

In July 2007, China State Forestry Administration, China National Petroleum Corporation (CNPC) and China Green Foundation jointly sponsored and established the China Green Carbon Fund (CGCF). They developed auditing standards of forestry carbon sequestration projects under this Fund, including eligibility criteria, additionality, baseline accounting standards, methodology, monitoring reports, etc., thereby achieving a forestry carbon sequestration, auditing and monitoring technology system. Based on this system, the CGCF supports forestry carbon sequestration projects in Beijing, Heilongjiang, Gansu, Hebei, Hubei, Zhejiang, Guangdong and other provinces; the approved carbon sequestration is up for sale in each market, with the price per tonne being approximately US$3. These projects have attracted enterprises, banks and individuals that hope to achieve carbon offsets and carbon neutralisation, with the current market situation featuring demand greater than supply.

Y. Zhu et al., *Transforming Rural Communities in China and Beyond*,
DOI 10.1007/978-3-319-11319-7_4

The demand for voluntary market forestry carbon sequestration outstrips supply mainly because, first, enhanced environmental awareness and social responsibility of Chinese enterprises, banks and individuals has increased demand in purchasing carbon sequestration. Second, although woodland eligibility requirements have been relaxed to some extent, China's local forestry carbon sequestration still has very strict requirements on additionality and baseline audit in order to guarantee the carbon reduction. Therefore, development of forestry carbon sequestration is not 'the icing on the cake' for farmers—forestry carbon sequestration doesn't add to the original economic benefits of forestry. In order to meet the audit requirements of carbon sequestration forests, under the farmers' woodland management model the choice of tree species must be limited, so not all farmers' carbon sequestration forest operating income could exceed that of farmers running economic forestland.

Whether participation in forest carbon sequestration projects can increase farmers' income is related to farmers' geographical location and the participation mode of woodland in such projects. Currently, Chinese farmers participate in woodland forestry carbon sequestration projects mainly in the following forms:

1. Forestland operation rights are still concentrated in the village collective but have not been decentralised to individual farmers, and the land is leased to forestry companies in the name of the village collective. Then, forestry companies operate carbon sequestration forest by means of leasing woodland. In this case, the farmers can gain the woodland rental and employment opportunities to participate in forest protection in principle of proximity, where farmers' increase in income depends on the comparison between opportunity cost of land management and land rental plus the associated job remuneration.
2. The forestry company gains shares in the name of the village collective, and the village and the forest company jointly operate carbon sequestration forest by contract.
3. Independently run carbon sequestration forest in the name of the collective village.
4. The village collective or farmers can directly sell forestland to forestry companies by listing it for sale in the Forestry Exchange. Through the transfer of forestry rights, farmers' real woodland management right can change from the resource into capital, but farmers lose woodland ownership after getting lump-sum compensation.
5. Farmers become shareholders by merging their own woodland with professional forestry development cooperatives and run the business of carbon sequestration forest.
6. Farmers lease the woodland assigned to them to forestry companies.
7. Farmers become shareholders by merging their own woodland with that allocated to the forestry company to develop joint carbon sequestration forest projects.

The ability for farmers to gain increased income from participation in forestry carbon sequestration projects is also constrained by geographical regions, woodland property circulation patterns, the cost and market price of forestry carbon sequestration, the government's supporting efforts, and other external factors. While such projects can pool social capital to support emissions reduction, the development of ecological forest, and alleviating farmers' poverty, the government plays a critical role in strengthening the support and management of forestry carbon sequestration projects to reduce the level of risk and to protect the interests and income of farmers. This chapter presents a comprehensive overview on the present situation of China's forestry carbon sequestration and the impact of woodland property rights reform on the development of forestry carbon sequestration projects. Through a comparison of projects in a number of provinces, such as Guangxi, Zhejiang and Yunnan, it analyses the income changes of farmers participating in those projects in different regions with different modes of woodland property rights.

4.2 China's Forestry Carbon Sequestration and Forestry Rights System Reform

4.2.1 Forestry Carbon Sequestration in China: Objectives

Forestry carbon sequestration refers to the process of forest ecosystems absorbing carbon dioxide from the atmosphere and fixing them in vegetation and soil and thereby reducing the concentration of atmospheric carbon dioxide (Li, Gong, & Zhang, 2006). To slow the release of carbon emissions through forestry carbon sequestration has not only great potential for a better environment but also a significant cost advantage (Benitez, McCallum, & Obersteiner, 2004). In view of its important role in climate change and the cost advantages, China undertakes forestry carbon sequestration as a response to climate change and proposes the appropriate action program and development goals. In 2007, China's National Climate Change Program put forward the objectives, principles, key fields and policy measures to tackle climate change, of which increasing forestry carbon sequestration is one of the key fields. On September 22, 2009, at the UN Summit on Climate Change (New York), President Hu Jintao clearly proposed the development goal of "energetically increasing forestry carbon sequestration, striving to realize forest area in 2020 can be 40 million ha^2 more than that in 2005, forest stock volume increases 1.3 billion m^3 compared with that in 2005".

The Chinese government is committed to the development of forest carbon sequestration projects mainly for the following purposes:

First, Tackle Climate Changes China faces growing international pressure to reduce emissions and is in urgent need of finding appropriate emission reduction strategies and methods. Implementation of energy conservation and emission reduction in industry has made good progress, but marginal emission reduction costs also increased with the progress of such reductions, while forestry carbon sequestration abatement costs are relatively lower. Thus, the Chinese government pays considerable attention to emission reduction potential of forestry carbon sequestration projects. Afforestation and reforestation carbon sequestration projects involves a lot of technical problems, including the determination of project baseline and additionality, measurement and verification of carbon sequestration, as well as non-persistence, leakage and the impact of the project on the socio-economic and environmental problems. However, the Chinese government is still determined to vigorously promote its development thanks to the maintenance function of forestry carbon sequestration projects on climate change and ecosystem. According to the estimates of Li and Yuan (2011), China still has more than 40 million ha^2 of barren hills and wasteland and a considerable amount of marginal land that can be used for afforestation. Although it is likely to become more and more difficult and expensive to undertake afforestation, there is still a large potential for development'.

Second, Further Promote Construction of Ecological Forests Confronted by serious environmental crisis, extreme lack of water resources and degradation of soil quality, China needs a lot of ecological forest to conserve water sources and soil. The Chinese government defines ecological forest as: forest that is built mainly for the purpose of reducing soil erosion and sand hazards, and other eco-efficiencies, including soil conservation forest, water conservation forest, wind-breaking and sand-fixing forest, and so on. There is significant difference between ecological forest and economic forest with the primary purpose of the production of fruits, edible oils, beverages, spices, industrial raw materials and herbs in forestland management and other aspects. Woodland Property Rights in China are classified into state-owned forest and collective forestland, and collective forestland occupies the majority of woodland area. Zhang (2001) claimed that China owns a total forest area of 128.5 million ha, the area of collective forestlands accounts for 55 %. Miao and West (2004) considered that collective forest area accounts for 60 % of the total forest area of the country. According to the results of the Sixth National Forest Resources Inventory, the collective forestland area occupies approximately 170.7 million ha, accounting for 60.37 % of national forestland area.

After the implementation of the rural land property rights system reform in China, the country has weakened control over collective woodland management; collective woodland management rights belong to the village community and local farmers, while the government can only stimulate and guide collective woodland operations through economic means but not force the implementation with executive orders. In this context, in order to vigorously develop ecological forest, the country must adopt related economic policy to give sufficient economic incentive to village communities and farmers in possession of operation and management rights of collective forestland. Although the Chinese government has allocated a large amount of financial subsidies, each household can only gain very little of the money assigned. Therefore, China's development of ecological forest is constrained by a serious lack of economic incentives. Within the market economy environment, it is difficult for the government to compel farmers to take the initiative to give up development of economic forests in favour of ecological forests. But due to the drinking water crisis in many provinces of China, as well as the serious decline of national soil quality, the Chinese government has to seek more economic policy instruments to stimulate the development of ecological forests. To achieve this end, the economic policy is pooling more resources to stimulate the development of forest carbon sequestration projects.

In China, forestry carbon sequestration is currently experiencing rapid development in the voluntary market; buyers are enterprises, individuals and social groups that have a passion for the protection of the ecological environment and hope to achieve carbon neutralisation. Under the promotion of the Chinese government, most purchasing funds are from farmers or forestry companies operating forest carbon sequestration projects, leading to an increase in the market for ecological forest.

Third, Increase Farmers' Income Farmers are the main suppliers of agricultural products, forestry products and ecological products. Under the conditions of a market economy, if farmers can only make a low income, they do not want to expand the supply of these products. In addition, in recent years, the gap between farmers' income and urban residents' income has gradually expanded, and the income gap between poor farmers and wealthy farmers is also expanding, leading to many social problems. Thus, the Chinese government is very concerned about the increase of farmers' income, especially the livelihoods of those in poverty-stricken regions, and the development of forestry carbon sequestration is an important mechanism to reduce poverty.

4.2.2 Types of Forestry Carbon Sequestration Transactions

CDM forestry carbon sequestration under the framework of the Kyoto Protocol requires that afforestation restores forest on land where there has been no forest during the past 50 years due to direct human intervention; reforestation means to restore forest on the land that has been unforested since December 31, 1989. This greatly limits the scope of qualified forestry project activities and the amount of eligible land that can be used to carry out CDM projects (China State Forestry Administration, 2012). As far as China is concerned, in order to meet these conditions mentioned above, the plantation land is mainly distributed in the northwest region of China (Zhang & Wu, 2005, p.116). These areas are remote and inaccessible, constrained by harsh natural conditions, posing difficulties in afforestation as well as high afforestation costs and monitoring costs, a long cycle of generating carbon sequestration and extremely limited amounts. These constraints have greatly restricted the development of forestry carbon sequestration projects under the Kyoto Protocol framework.

In terms of implementation of Kyoto Protocol projects, China has established a number of pilot projects and signed carbon emission reduction contracts with more than a dozen developed countries in the world. The China Guangxi Pearl River Basin Reforestation Management Project absorbs large amounts of CO_2 through planting of Masson pine, sweet gum, Griffith oak, schimacrenata, and eucalyptus in mixed forest. The World Bank Bio Carbon Fund will purchase CO_2 equivalent at a price of US$4.35 per tonne, and is expected to purchase 480,000 t over 15 years. Meanwhile, the implementation of the project provides migratory corridors and habitats for wild fauna and flora in surrounding nature reserves, thereby protecting biodiversity and controlling soil erosion in the project area. In addition, this project will provide tens of thousands of temporary jobs for local farmers and create 40 long-term employment opportunities, and 5,000 farmers will benefit from the sale of carbon sequestration, as well as wood and non-wood forest products (Forestry Bureau of Guangxi Zhuang Autonomous Region, 2007). Such experience gained through the Pearl River Basin project led to the Northwest Guangxi CDM Reforestation Project being approved by the State Council in October 2008 (Li, Yang, & Chen, 2010).

By 2011, the China National Development and Reform Commission had approved a total of 2,597 CDM projects, including four forestry carbon sequestration projects, and expected annual carbon sequestration removals of 117,400 t of CO_2 equivalent. Each item is detailed in Table 4.1.

However, CDM afforestation or reforestation projects under the Kyoto Protocol framework face great obstacles in the implementation process, including demanding requirements on the eligibility of the land, serious impact of the solution to current non-permanent problem on Certified Emission Reduction (CER) prices and the interest of the purchaser, and an overly complex project preparation and registration process. In April, 2012, there were only 37 developed and registered CDM afforestation or reforestation projects in the world, but emission reduction projects were up to 4,028. Therefore, the Chinese government had to shift its focus to non-Kyoto carbon sequestration projects.

In terms of non-Kyoto Protocol carbon sequestration projects, forestry projects in the voluntary market hold a dominant position. The demand for such carbon sequestration in China mainly comes from fulfilment of enterprises' social and environmental responsibility, as well as personal carbon footprint offsets. There is a huge market demand, which is growing along with the increased knowledge and concerns about environmental protection. However, the current market price is

Table 4.1 CDM carbon sequestration project approved by NDRC (by July 30, 2011)

Approval numbers	Project title	Location	Foreign cooperators	Estimate annual reduction
42	China Guangxi Pearl River basin reforestation management project	Guangxi	Italy, Spain	20,000
1,513	China Northwest of Sichuan province degraded land afforestation and reforestation project	Sichuan	(Unilateral project)	26,000
1,643	China Guangxi Northwest degraded land reforestation project	Guangxi	(World Bank)	70,272
1,847	China Liaoning Kangping desertification prevention small-scale afforestation project	Liaoning	Japan (Keio)	1,124
	Total			117,396

Source: National Development and Reform Commission, Department of Climate Change website (Clean Development Mechanism website) http://cdm-en.ccchina.gov.cn/

rather low, about US$3 per tonne, but the market still has great potential and larger appreciation room (see Table 4.2, East China Forestry Exchange, carbon sequestration trade on November 1, 2011).

In order to promote the development of forestry carbon sequestration in a voluntary market, in July 2007, the State Forestry Administration, China Green Foundation and CNPC jointly initiated the CGCF to support forestry departments to tackle climate change action, as was explained in Chap. 3. The fund establishes a platform for enterprises for investment in afforestation and acquisition of CO_2 emission rights, storing carbon credits in advance. This may not only help enterprises to voluntarily reduce carbon emission at a lower cost, but also establish a good public image, increase forest vegetation and consolidate national ecological security.

On April 4, 2008, Beijing Fangshan District and Badaling Carbon Sequestration Afforestation projects were officially launched by CGCF. The CGCF Qinglonghu town carbon sequestration afforestation project in Fangshan District in Beijing is funded by CNPC with the investment of 3 million yuan, to join hands with Qinglonghu Town in Fangshan District to create a 400 ha^2 carbon sequestration forest in the following 20 years. On completion, this project will not only increase plant species and biodiversity in the project area, but also improve the local ecological environment. It is estimated that this project can annually absorb nearly 4,000 t CO_2. In 2007, other carbon sequestration projects were implemented in Shuangjiang, Linxiang, and Gengma Counties and other places in Yunnan province, fully invested by Canada Sino-Forest Corporation for the reforestation.

In order to standardise voluntary market trade of forestry carbon sequestration, the China State Forestry Administration Climate Change and Energy Conservation Emission Reduction Leading Group Office compiled the CGCF Guide for Carbon Sequestration Project Measurement and Monitoring. This guide not only fully adheres to IPCC Guidelines and methods of measurement of forestry carbon sequestration stocks in other voluntary carbon markets in the world, but also considers the reality of China Forestry. As China's first guideline on carbon sequestration projects measuring and monitoring in a voluntary carbon market, it not only applies to the CGCF, but can also be used for carbon measurement and monitoring of other domestic afforestation projects.

In order to explore the road of China's forestry voluntary carbon market complying with international market standards, China implemented Forest Carbon Sequestration Biodiversity Projects in Sichuan, Yunnan and other places, which became the first such projects in the world to obtain certification of multi-efficiency climate, community and biodiversity standards (CCBS). CCBS was developed by the Climate, Community and Biodiversity Alliance (CCBA[1]) to guarantee that the project at the design stage will be able to give equal attention to the benefits of climate, community and biodiversity. Since 2004, the CCBA, China Forestry Administration, The Nature Conservancy (TNC), 3 M Company and UTC have cooperated in planting and demonstrating multi-benefit forests in mountainous biodiversity hotspot areas in south west China, such as northern Sichuan and western Yunnan, and selling carbon credits through the market mechanism after third-party certification. In addition, some parts of the projects were certified under CDM. The

[1] http://www.climate-standards.org/category/projects/

Table 4.2 East China Forestry Exchange, carbon sequestration trade on November 1, 2011

Contract number	Name of the subject matter	Type of the subject matter	Transferor	Transferee
HDT2011000001	Authorised by the estimate of carbon credit 10,000 t of carbon dioxide equivalent	Forestry carbon sink	CGCF	Alibaba
HDT2011000002	Authorised by the estimate of carbon credit 10,000 t of carbon dioxide equivalent	Forestry carbon sink	CGCF	Ge Shan construction group
HDT2011000003	Authorised by the estimate of carbon credit 10,000 t of carbon dioxide equivalent	Forestry carbon sink	CGCF	Dezheng Zhiyuan carbon fund (Tianjin Dezheng Zhiyuan equity investment management cooperation enterprise)
HDT2011000004	Authorised by the estimate of carbon credit 10,000 t of carbon dioxide equivalent	Forestry carbon sink	CGCF	Kai xuan Street, Jiang gan District, Hangzhou
HDT2011000005	Authorised by the estimate of carbon credit 10,000 t of carbon dioxide equivalent	Forestry carbon sink	CGCF	Hangzhou Qianwang accounting firm (Shen li)
HDT2011000006	Authorised by the estimate of carbon credit 10,000 t of carbon dioxide equivalent	Forestry carbon sink	CGCF	Fuyang wood comprehensive market
HDT2011000007	Authorised by the estimate of carbon credit 10,000 t of carbon dioxide equivalent	Forestry carbon sink	CGCF	Longyou foreign trade of bamboo shoot factory
HDT2011000008	Authorised by the estimate of carbon credit 10,000 t of carbon dioxide equivalent	Forestry carbon sink	CGCF	Hangzhou Hongda Office Furniture Manufacturing Co. Ltd
HDT2011000009	Authorised by the estimate of carbon credit 10,000 t of carbon dioxide equivalent	Forestry carbon sink	CGCF	Zhejiang Mulaolao Toy Handicraft Co. Ltd.
HDT2011000010	Authorised by the estimate of carbon credit 10,000 t of carbon dioxide equivalent	Forestry carbon sink	CGCF	Hangzhou Yuyue Investment (Hangzhou Yuyue Commercial Real Estate Investment Management Co. Ltd)
HDT2011000011	Authorised by the estimate of carbon credit 10,000 t of carbon dioxide equivalent	Forestry carbon sink	CGCF	Zhejiang ShinDai Sanyuan Holdings Limited
HDT2011000012	Authorised by the estimate of carbon credit 10,000 t of carbon dioxide equivalent	Forestry carbon sink	CGCF	Xia Chunhong

Means of exchange	Unit price (Yuan/ tCO2e)	Knock down price (Yuan)	Method of payment	Carbon afforestation project	Crediting period
Negotiating transfer	18	180,000	Installment. The first payment was 60 % of total cost	CGCF Guangdong Longchuan carbon afforestation project (57254tCO2e)	20 years (2008-01-01 to 2027-12-31)
Negotiating transfer	18	180,000	Installment. The first payment was 60 % of total cost	CGCF Guangdong Shantou Chaoyang carbon afforestation project (60610tCO2e)	
Negotiating transfer	18	180,000	Installment. The first payment was 60 % of total cost	CGCF Gansu Dingxi Anding carbon afforestation project (4300tCO2e)	
Negotiating transfer	18	180,000	Installment. The first payment was 60 % of total cost	CGCF Zhejiang Lin'an moso bamboo carbon afforestation project (8155tCO2e)	
Negotiating transfer	18	180,000	Installment. The first payment was 60 % of total cost	CGCF Beijing Fangshan carbon afforestation project (6495tCO2e)	
Negotiating transfer	18	180,000	Installment. The first payment was 60 % of total cost	CGCF Gansu Qingyang Heshui state-run forestry field carbon afforestation project (11757tCO2e)	
Negotiating transfer	18	180,000	Installment. The first payment was 60 % of total cost	148,571 t of carbon dioxide equivalent in total	
Negotiating transfer	18	180,000	Installment. The first payment was 60 % of total cost		
Negotiating transfer	18	180,000	Installment. The first payment was 60 % of total cost		
negotiating transfer	18	180,000	Installment. The first payment was 60 % of total cost.		
Negotiating transfer	18	180,000	Installment. The first payment was 60 % of total cost		
Negotiating transfer	18	180,000	Installment. The first payment was 60 % of total cost		

Yunnan Tengchong small-scale reforestation project, developed via a multi-benefit forest project, obtained the world's first CCBA Gold Certification in January, 2007.

4.2.3 Forestry Carbon Sequestration Trade and Transfer of Forestland Use Rights

Forestry carbon sequestration projects require a sound woodland property rights registration system, competent asset evaluation institutions and perfect woodland evaluation system. Meanwhile, projects require intensive management of forestland, but farmers of small-scale decentralised management cannot become the project owner. For example, the CGCF Guide for Carbon Sequestration Project Measurement and Monitoring requires that forestry carbon sequestration projects should have a certain scale; in principle, each project shall be no less than 5,000 acres in the implementation scale and at least 20 years old at the entire implementation period. Project implementation units must take effective measures to ensure that the afforestation project achievements can be accompanied by reasonable management and protection, and can be included in the unified management of local forest resources within 20 years after afforestation (China State Forestry Administration, 2009). In addition, the CGCF Guide emphasised that, in regard to land eligibility auditing, afforestation land ownership must be sufficiently clear, with land use rights certificates issued by local government departments. Obviously, forest carbon sequestration projects require clear forestland property rights, and need to be based upon a certain scale of profitable land.

China's original collective forestland property is not clear enough, which has seriously affected the implementation of carbon sequestration projects. Such unclear land property rights include unclear woodland management rights and unclear 'four boundaries' of woodland.

Since the founding of the People's Republic of China in 1949, China's forest tenure system has undergone several major changes. Before 1981, China's woodland was made up of state-owned and collective forestland. Collective forestland was owned by the people's communes and under the unified management of the collective. However, the people's communes only had the right of collective cultivation but not the rights of leasing and trading forestland, and even the distribution of forest products were still subject to layers of constraints of higher levels of government. The national government strictly controlled the operation of collective forestland through the administrative system. Therefore, the collective forestland owned by the people's communes had incomplete property rights.

In 1981, China began to implement a household contract responsibility system in rural areas; collective forestland originally owned by the people's communes was also contracted to farmers. As the farmers were doubtful about the long-term stability of the policy of assigning forest to each household, and worried that the woodland would be taken back by the government in the future, many farmers cut down a lot of trees, leading to a serious loss of forest resources.

In order to reverse the deforestation, promote the forestry carbon sequestration trade and meanwhile increase farmers' income, the Chinese government initiated a new round of collective forestry right system reform. The core elements of reform were: first, clarify and stabilise farmers' forestland use rights. The Chinese government extends the farmers' woodland contracting period to 70 years and implements field boundary surveying and registration of farmers' woodland allocation, and issues the national uniform Forest Tenure Certificate. Second, allow the transfer of forestland use rights of farmers to achieve intensive and scale management of woodland. The revised Forest Law of the People's Republic of China establishes the forestland use rights transfer system, so that forestland use rights become independent civil rights. Forestland use rights contain forests, trees and other above-ground resources, as well as the right to use, gain income from and manage forestland. In order to ensure that the circulation of such rights cannot cause the forestland to be converted to other use, Land Administration Law and Forest Law expressly stipulate that the transfer of land use rights shall not change the forestland use. In order to make the transfer of forestland use rights have more of a legal basis, in 2007, section 133 of the Property Law provided as follows: "Wasteland and other rural lands can be contracted through bidding, auction and public consultation, and other means". In accordance with the Law of Rural Land Contract and the relevant provisions of State Council, land contract and management rights can be transferred through sale, shares, mortgages or other modes. Under the protection of these legislations, forestland use rights become a resource property right of farmers.

Forestland use rights circulation modes mainly include:

First, Forestry Rights Trading The trading of forest rights means that forestry right owners will transfer all or part of their forestland use rights to others. Farmers mainly take the following measures in forest right trading:

(a) private negotiation between both trading sides
(b) register in forest rights trading centre and trade publicly
(c) private intermediary organisations coordinating the trade
(d) trade by collective contract.

According to a sample survey on forestland use rights of farmers in Yichun City, Jiangxi Province, 52 % of these farmers will choose option (a) private negotiation, which is basically suitable for small-scale forestland use rights trading between farmers. Farmers are very different in their household types and side-industry business conditions as well as interest in woodland operations and management capacity. Different farmers have different expectation on woodland operating income, so there are many small-scale forestland use rights trading between farmers. Farmers choose private negotiation mainly because, first, their circulating forestland is on a small scale and cannot be traded in the forestry exchange center. Second, the parties in the trade are very familiar with each other and may even be in the same village, under the protection of village customs. It is also safer for them to choose private negotiation because of its reduced transaction cost.

Among these farmers, nearly 41 % choose the government-run forestry exchange for registration, assessment and public trading. Trade in the forestry exchange involves complex procedures. However, if the woodland to be traded is on a large-scale, farmers would think it is safer to choose the exchange. Despite the complex procedures, the trading can be guaranteed by laws and regulations (see Table 4.3, list of part of forest rights trading items from 2012 to now of Huadong Forestry Exchange).

Table 4.3 List of part of forest rights trading items in 2011

	Name of the subject matter	Ownership property	Land transfer area (Mu)	Type of the subject matter	Listed price	Transaction price	Premium rate	Listing date	Transaction date	A deal, or not?
1	Yunnan Dali Jianchuan county 200,000 Mu forest tenure transfer	Collective	200,000	Forest tenure						N
2	Lishui Jingning Dajun Fuye village (Niucao Lake) Mangnolia officinalis forest transfer of the contract for the managerial right	Collective	34,000	Forest tenure						N
3	Jingning Dajun Fuye village (Zhuer Lake) Mangnolia officinalis forest Transfer of the Contract for the managerial right	State-owned	34000	Forest tenure						N
4	Jinyun Kuocang hill Provincial forest park Transfer of the Contract for the managerial right	State-owned	17,279	Forest tenure						N
5	Jiangxi Luxi county Wanlong, Shangbu 15,000 Mu moso bamboo transfer of the management right	Enterprise	15,000	Forest tenure						N

6	Shanxi somewhere about 13,000 Mu forest tenure transfer	Natural person	13,000	Forest tenure						N
7	Songyang county Sidu village 5,870 Mu torreya grandis health garden transfer of the contract for the managerial right	Enterprise	5,870	Forest tenure						N
8	Sichuan Liangshan Yanyuan county Gaizu village 5,054.8 Mu timber forest tenure transfer	Natural person	5,054.8	Forest tenure						N
9	Yunnan Pu'er Jingdong Yi autonomous county Huashan Yunpan village personal forest tenure transfer	Natural person	5,049	Forest tenure						N
10	Fuyang Lishanhe village "Dapan hill" 2,335 Mu forest tenure transfer	State-owned	2,335	Forest tenure	1,014.816	2,501.93	1.4654026	2011/3/22	2011/4/22	Y
11	Pujiang county Qianxi village 2,113 Mu forest tenure transfer	Collective	2,113	Forest tenure	140	288.88	1.0634286	2011/11/1	2011/11/25	Y
12	Kaihua county Chihuai town Batou village 2,045 Mu Forest tenure transfer	Collective	2,045	Forest tenure	90	90	0	2011/7/1	2011/7/7	Y

(continued)

Table 4.3 (continued)

	Name of the subject matter	Ownership property	Land transfer area (Mu)	Type of the subject matter	Listed price	Transaction price	Premium rate	Listing date	Transaction date	A deal, or not?
13	Fuyang Qiaotou town Sanxi village sugarbush 1,973.5 Mu Forest tenure transfer	Collective	1,973.5	Forest tenure	44	105	1.3863636	2010/9/29	2010/10/26	Y
14	Dongyang Linggangu village 1,181 Mu forest tenure transfer	Collective	1,181	Forest tenure	138	142.8	0.0347826	2011/5/15	2011/6/10	Y
15	Fuyang Qiaotou town Sanxi village sugarbush 1,086 Mu forest tenure transfer	Collective	1,086	Forest tenure	18	55	2.0555556	2010/9/29	2010/10/26	Y
16	Fuyang Qiaotou town Sanxi village Xiabi east 1,058 Mu forest tenure transfer	Collective	1,058	Forest tenure	120	218	0.8166667	2010/9/29	2010/10/26	Y

17	Kaihua Suzhuang town Fangpo village 3rd group "Shiwu" 32 Mu forest tenure transfer	Collective	32	Forest tenure	18	20	0.1111111	2011/8/10	2011/8/18	Y
18	Lin'an Daoshi town 30 Mu hickory forest tenure transfer	Natural person	30	Forest tenure	45	48	0.0666667	2011/11/17	2011/11/30	Y
19	Kaihua Suzhuang town Fangpo village 1st group "Changmaowu" 29 Mu forest tenure transfer	Collective	39	Forest tenure	20	20.2	0.01	2011/8/10	2011/8/18	Y
20	Kaihua county Changhong Hongqiao village "Qianwu" 20 Mu forest tenure transfer	Collective	20	Forest tenure	21	22	0.047619	2011/11/15	2011/12/6	Y

Note: 1–9: Forest tenure transaction projects whose land transfer areas are all more than 5,000 Mu, however, there is no one succeeding in transaction
10–16: Forest tenure transaction projects whose land transfer areas are all more than 1,000 Mu and succeeding in transaction, 7 in total, have all been listed
17–20: List 4 successful forest tenure transaction projects whose land transfer areas are the smallest

In the sample areas surveyed, only 5 % of farmers choose the trade coordinated transactions by private intermediary organisation, while 2 % of farming households choose trade via the village collective. Most farmers are wary of choosing the latter, because they consider village cadres untrustworthy and fear they will damage their interests.

Second, Forestry Property Investment as a Shareholder Forestry property shareholders turn farmers' forestland use rights into capital and assemble them so as to form a management scale and resources industrial base in a short time period. Forest ownership shares are mainly in the form of foresters' joint shareholding cooperative forest or company plus farmer. Foresters' joint-stock cooperative forest can improve the forestry output efficiency. For example, in Jiashang village, Nakou Town, Shaowu City, Fujian Province, the villagers own 0.575 ha^2 per capita of the woodland area. If they had chosen household management, woodlands were rather scattered. In order to achieve large-scale operation of the woodland, villagers joined together to form a joint-stock cooperative forest. They conducted an assessment on 175.7 ha^2 of forestland in the village, with the assessment price being RMB 1.386 million. Then 35 farm households in the village were divided into 35 individual stocks, while villagers became shareholders through forestland use rights and thus formed a joint-stock cooperative forest. Meanwhile, they also developed the Charter of Jiashang Villagers' Joint-stock Cooperative Forest and carried out production and operation in accordance with the principle of "paid shares, co-management, guaranteed rights and interests, profit-based bonus".

A number of enterprises and individual farmers have achieved better outcomes in some places. For example, in Siqian Township, Guangze County, Fujian, by adopting the mode of 'company + farmer + basement', local farmers became shareholders through their own bamboo forests and formed a community of interests with a local company; the company will invest RMB 900 per ha^2 each year, including forestland use fees of RMB 300 per ha^2, phyllostochys pubescens forest tending fee RMB 300 per ha^2, and assumed responsibility for the creation of a tractor road. They will share the profit of harvested phyllostochys pubescens 50:50 (Yang, 2008).

Third, Leasing Forestry Property Rights Forestry property rights leasing means the transfer of forestland use rights for a certain period of time by charging rent, interest shares or other forms. Usually, some powerful enterprises, institutions, foreign merchants or persons with financial capability use their own capital, technology, and information advantages to lease all the forestlands in decentralised management of farmers. The lease term typically runs for 30–50 years, while a few of them can be as long as 70 years. Rents are mostly paid in cash, but a few of them are paid at one time or several times. The lessee is responsible for all business activities after leasing, with risks and profits also belonging to the lessee.

Fourth, Forestry Right Mortgage Forestry Right Mortgage means the owners of forestland use rights obtain loans from the bank by mortgaging the forestland use rights recorded on their Forest Tenure Certificates. When the farmers who mortgage the forestry property rights fail to pay due debts and interests, the bank can sell or auction the collateral to settle their claims.

Both forestry carbon sequestration project operation and other forms of forestry management need financial support. Forest tenure mortgages increase financing channels for foresters in poor mountainous areas, so it is a good way to increase forestry resources, promote forestry carbon sequestration, and alleviate the poverty of farmers in these areas. Therefore, the Forest Right Mortgage Loan wins greater attention from the Chinese government, the People's Bank and the FDA, which jointly issue various documents to promote the popularity of the Forest Right Mortgage Loan in China.

4.3 The Influence of Forestry Carbon Sequestration Trading and Forestry Rights Transfer on Farmers' Livelihood

For farmers in mountainous areas, forestland is the most basic resource foundation for their survival (Boyd, Gutierrez, & Chang, 2007). Thus, the forestland property rights system and forestry carbon sequestration trading system implemented by the Chinese government will inevitably affect farmers' income in mountainous areas and their progression of development. Both the carbon sequestration trading system and forestry tenure reform established, as one of their system goals, to increase farmers' incomes and improve farmers' welfare. Circulation of forest rights can promote the centralised and large-scale operation of the forestland, clearing the way for the implementation of carbon sequestration projects. However, such projects have very complex influences on farmers and rural communities.

In theory, a forestry carbon sequestration project that is designed perfectly should not only be able to raise the level of farmers' livelihoods by improving farmers' control and management right to forest resources, but also could address climate change by reducing greenhouse gas emissions. Moreover, a lot of people misunderstand these projects; they believe that their implementation is bound to increase the income of farmers, because of the income from both carbon sequestration the farmers' original forestry income.

However, forest carbon sequestration projects have many stringent requirements. The implementation of projects on a patch of forestland does not simply mean the original forestry income plus carbon sequestration income for farmers, because stringent requirements on carbon sequestration projects will limit or change foresters' traditional business model of forestland and increase their afforestation costs. Unless forestry carbon sequestration income can make up or exceed the cost of the new afforestation, foresters' implementation of forestry carbon sequestration projects may reduce their income.

First, a forestry carbon sequestration project requires that forestland management activities must reduce carbon emissions to the maximum extent, which may increase foresters' land management costs. Carbon sequestration must be the primary purpose, but there will be some difference in land use compared with conventional forestland management and some special requirements in order to maximise carbon sequestration. For example, China Green Carbon Fund Interim Provisions on Carbon Sequestration Projects Afforestation Technology points out that the implementation of a project should minimise or avoid greenhouse gas emissions resulting from the project activities. Emissions of these activities may come from use of machinery for land preparation, motor vehicles to transport seedlings and fertiliser. In other words, in order to reduce carbon emissions in woodland management, farmers are required to use as few mechanical operations as possible while adopting manual operations as much as possible, and it is preferable to avoid use of chemical fertilisers.

Second, special provisions on land preparation technology may make some economic forests incapable of being deeply plowed and planted, thus affecting the income of foresters. In the calculation of new carbon sequestration of a forest ecosystem, it is required to first approve a baseline carbon sequestration amount; carbon sequestration that is available for trading is the increased sequestration above the baseline level. Therefore, the forestland management model—based on carbon sequestration goals—stresses that the process of land preparation and afforestation shall minimise disturbance to the original vegetation carbon and soil carbon. As a result, we need to choose ribbon or massive land preparation but avoid deep harrowing. Technical requirements for carbon sequestration afforestation stipulate land preparation techniques, requiring that counter-slope faced pit soil preparation, strip land preparation and fish-scale pit land preparation shall be adopted in most cases. Counter-slope faced pit soil preparation is suitable for slopes above 25° (inclusive), the specification should not be less than 40×40 cm. Slopes below 25° shall generally adopt strip land preparation, which shall be carried out along the contour, while the strip width shall be dependent on the spacing between rows in the afforestation, generally 40–100 cm, with the depth of 30–50 cm, and the length determined by the terrain, but should not exceed 10 m, while the strip shall maintain a horizontal direction along the contour. Fish-scale pit land preparation is applicable to sloping land in arid, semi-arid areas. Fish-scale pit length extends along the direction of the contour, usually 60–80 cm, the short diameter is slightly shorter than long diameter, and the depth shall be 40 cm or above. Such strict land preparation technical specifications will limit the variety of trees or obstruct the planting of understory crops, because the understory crops generally require deep plowing. Foresters lose the revenue from understory crops, and their total revenue from management of carbon sequestration forests will be inevitably affected.

Third, forestry carbon sequestration projects must meet additionality stipulations. In order to meet these requirements, the compensation standards of various projects constrain the choice of forestland, so not all forestland operations can get carbon sequestration compensation. Only afforestation on long-abandoned woodland can meet the requirements of additionality of forestry carbon sequestration projects. As explained earlier, CDM projects have the most demanding restrictions on forestland. On the other hand, in terms of voluntary carbon market projects, the CGCF has relatively looser restrictions in terms of additionality provisions than CDM projects, but it is also difficult to meet the standard. Its technical stipulation clearly defines the project site selection to verify its additionality. The technical stipulation requires that the afforestation land implementing carbon sequestration projects must be the land that has been unforested before or since January 1, 2000; and the forestland must have no commercial competitiveness, meet certain technical difficulty of afforestation, and not have natural regeneration capability. Forestland that can meet the provisions of additionality inevitably has higher operating costs than general forestland, but the output is not necessarily higher; so if there is no adequate carbon sequestration compensation, it is difficult for foresters to get better returns from such kinds of land operation.

Fourth, requirements generally stipulate that a carbon sequestration project period is 20 years or more. In the crediting period of such a project, we must ensure that project deliverables can be maintained properly. It is required to strengthen the management and protection for new afforestation and immature forestland, so long-term measures must be taken to avoid damage caused by people and livestock in addition to routine weeding and tending. Based on IPCC 2006, carbon pooling is a system used for the measurement of carbon sequestration removals, including aboveground biomass, underground biomass, dead wood, litter materials and soil organic carbon. According to the biomass calculation method, any behaviour causing impact on the biomass will be restricted. In order to make the biomass free from the impact of non-project factors, farmers have no right to graze stock in the project woodland, cut down branches or even plant underground crops.

For the above reasons, when foresters participate in forestry carbon sequestration projects, the impact on their incomes greatly varies in different areas depending on the forestry right participation mode, and the price of carbon sequestration.

4.4 Afforestation for Carbon Sequestration in Remote Regions

In general, for farmers in mountainous areas remote from the market, their forestland has to be have been abandoned for more than 10 years to comply with the additionality requirements of carbon sequestration projects. Even if approved for domestic voluntary market carbon sequestration in China, this kind of land generally has higher afforestation costs. If they do not resort to external support, farmers can only get an extremely small income from their forestry rights because opportunity cost of the operation of such land is almost zero. But if foresters can partner with an afforestation company and win government support, they can get increased gains.

If foresters can lease forestland to an afforestation company or become shareholders through the forestland to jointly develop and operate a carbon sequestration project, all funds are invested by the company. Farmers and the company get bonuses proportionally from the carbon sequestration income and timber sales income. These kinds of projects can improve the income of farmers. Farmers can get woodland rent, salary for participation in afforestation and forest management, and can also share the tree sales revenue and carbon sequestration income if they become shareholders. In remote mountainous areas, where a patch of woodland has been abandoned for 10 years, it is difficult for farmers to have enough capital and technology to afforest on this wasteland even if they own the forestry rights. After participation in carbon sequestration projects, farmers can get increased revenue including rent, timber sales income, carbon sequestration income, and wages for participation in afforestation and forest management. By such means, in remote mountainous areas away from the market, the implementation of forest carbon sequestration projects can increase revenue for foresters.

However, so as to maximise their own interest, an afforestation company always hopes to lease woodland at a lower price or allocate fewer benefits to farmers participating in projects as shareholders of woodland investment. In this case, the foresters' increased income and the community's welfare improvement rely on the negotiations between foresters and the company. Single individual farmers have weaker negotiating ability. On the other hand, strengthening the community collective ownership property of the forestland in the negotiations can enhance farmers' cohesion and guarantee that farmers benefit from carbon sequestration projects. In addition, another benefit of community collective negotiation with an afforestation company is that improving community infrastructure can be a content of the negotiation. Remote mountainous areas generally have relatively lower levels of infrastructure; farmers in the community generally have lower income, so they have no spare money to develop community economy, and a village collective has no funds to be redeployed, thus community infrastructure is allocated no investment. If individual farmers negotiate with a company, due to the public goods characteristics of community infrastructure, no farmers will mention building community infrastructure as the contents of the negotiation of the project revenue. Only by means of strengthening the collective ownership property of the woodland, and negotiating in the name of village collective, can farmers include community infrastructure into the content of the negotiations.

However, this mode must avoid the corruption of village leaders, that is, those who sacrifice the interests of the village collective to cooperate with an afforestation company at a lower cost, and in return receive bribes. Such corruption and bribery will damage the community interests and affect the income of farmers participating in carbon sequestration projects. Therefore, an open and transparent information and monitoring mechanism is very important in the negotiation.

4.5 Case Studies

4.5.1 Case One: Guangxi Pearl River Basin Reforestation Project

The first forestry CDM project developed in China, Guangxi Pearl River Basin Reforestation Project, implemented in Cangwu and Huanjiang Counties, is a project of this type. Launched in 2006, the project cycle will last for 30 years, with the goal to afforest 4,000 ha^2 in Cangwu and Huanjiang Counties; all project woodlands are non-forested land since December 31, 1989 (UNFCCC, 2006). According to the project design documents, the BioCarbon Fund will provide US$2 million to buy 48 t of CER incurred from the project, at the price of US$4.35 per tonne, with CER generated in the eventual project to be purchased by Italy.

Cangwu and Huanjiang Counties are poor mountainous areas in China's south west, where there is a low degree of marketisation. In 2005, the annual per capita income of farmers was US$267 in Cangwu County and US$245 in Huanjiang County. Residential areas of ethnic minorities are located here; Huanjiang County is the autonomous county of Maonan ethnic minority and is also categorised as a poverty-stricken. Mountainous areas account for more than 70 % of the total area of the county. Water, electricity, roads and other infrastructure is still backward, mountain dwellers are poor and live in dispersed places, and many children in mountainous areas are stricken by malnutrition as the result of poverty.

In Huanjiang County, farmers participate in carbon sequestration projects mainly in the form of the village collective as the negotiator to run stock cooperative afforestation with an afforestation company. Farmers become shareholders through forestry rights investment, and the afforestation company is responsible for investment, providing technology, management and bearing the natural disaster risk and investment risk. The village collective signs a contract with the company to clarify the afforestation management responsibility, investment and share of proceeds. Revenue share proportion: 40 % of net income of forest products, 60 % of CER sales income shall be owned by local farmers; 60 % of net income forest products, 40 % of CER sales income shall be owned by afforestation company. The company will give employment priority to local farmers to participate in land preparation, reforestation and management and protection activities, and pay farmers salaries for their labour inputs so as to ensure their short-term economic income. This management mode benefits 4815 farmers; according to estimates, during the project implementation, these farmers get an increase in income of US$18.2 million from their participation in carbon sequestration projects, including about US$15.6 million of employment income (a temporary worker's income is about US$3 per labour day, and a long-term worker's income is US$900 per person per year). Project areas in Huanjiang are concentrated in ethnic minorities, all employment opportunities are provided for local ethnic minority groups), 40 % of forest products income is US$1.4 million, and 60 % of carbon sequestration income is US$1,200,000. In 2006, these farmers gained an income increase of US$34 per person annually.

In addition, Huanjiang County is a remote mountainous area, and the project site has very poor transport conditions, with arduous mountain roads constituting an important reason for villagers' poverty. After the implementation of the carbon sequestration project, the village collective proposed that the afforestation company should assist the village community in building roads as one of the negotiation contents, while the company also needed to address transportation difficulties in order to better manage forest areas and communicate with the outside world markets. Thus, the implementation of the project brings the local community a good welfare improvement, i.e., building roads to enable villagers to communicate and trade more conveniently with the outside world.

In the implementation process of forestry carbon sequestration projects, the reason the local communities and farmers' income can be guaranteed and the project can be carried out smoothly is closely related to the fact that forestland property still belongs to the village collective and has not yet been allocated to farmers. However, along with the appearance of project revenue, communities and farmers are required to further clarify woodland property rights, bringing great difficulties to the implementation of the project.

In the early stage of the project design, the farmers agreed not to allocate forestland to each household, because the forestland of the project is a barren hill over 50 years old and thus farmers believed such forestland could not generate any income, so they would not further clarify the requirements of the Woodland Property Rights. Village collectives also failed to carefully delineate their respective forestlands; four boundaries were determined in accordance with forestry 'three definitions' in the early 1980s.[2] The four boundaries of forestland of each village collective have not yet been plotted on a topographic map but merely described in simple text. As a result, the four boundaries are not clear enough. Prior to the implementation of the project, the land was long-term barren and unforested, which had no embodiment of potential benefits, so there was no dispute concerning unclear forestland property rights between either village collectives or farmers.

However, along with the appearance of interests after the development of forestry carbon sequestration projects, foresters became aware of potential forestland income and villages started to dispute their respective forestland boundary. Consequently, 147.5 ha^2 forestland designed in the project cannot be implemented for afforestation due to land boundary disputes. In addition, at the beginning of the project implementation, forestland contract and management rights were owned by the village collective. With the development of the project proceeds, farmers began to request that the forestland originally in unified management of the collective should be reallocated to each farmer household. After forestland management rights are allocated to farmers, farmers must re-sign the forestland contract that was involved in the project and originally signed with the collective. Each farmer has different information acquired and a different understanding of the forestry carbon sequestration project, resulting in large transaction costs in re-signing the contract. In addition, some farmers took the opportunity to propose unreasonable demands or breach of contract, leading to 173.3 ha^2 of forestland becoming unavailable for the continued implementation of the carbon sequestration project.

[2] See Huanjiang Autonomous County Annals Compilation Committee, 2001, and Cangwu County Annals Compilation Committee, 1996.

4.5.2 *Case Two: Zhejiang Lin'an Bamboo Carbon Sequestration Project*

Whilst it was rather early to develop forest carbon sinks in Zhejiang Province, especially in Lin'an, there was a summit forum concerning China forest carbon sinks and development of low-carbon economics in Lin'an on October 27, 2010. During the forum, Lin'an was designated as the first forest carbon pilot site throughout the country, since Lin'an has its unique natural advantages—abundant forest resources, two national nature reserves, one national forest park, over 1,800 tree species, many of these tree species suitable for developing carbon forest sinks. Twenty-one mayors attending the forum also issued and signed the Lin'an Declaration, which is aimed at development of carbon forest and low-carbon economics. There was a workshop about forest carbon at the Doha Climate Summit, which enhanced the importance of bamboo forest carbon, and there were two highlights. Firstly, the release of the Bamboo Carbon Afforestation Methodology in China, which is the first and a world leading methodology in China; secondly, the International Network for Bamboo and Rattan (INBAR), CGCF, Zhejiang A & F University and People's Government of Zhejiang Anji signed an agreement called Construction of Bamboo Carbon Test Demonstration Area.

The resulting bamboo carbon sequestration project in Lin'an City, Zhejiang Province is a project supported by the CGCF. It belongs to the non-Kyoto Protocol project category and chooses 5-year barren forest as the project forestland. The afforestation project started land preparation in September 2007, and afforested phyllostochys pubescens carbon sequestration forest of 86.7 ha^2 respectively in Yanjia Village and Songxi village in Zaoxi Town, Lin'an City in 2008 and 2009. There are 700 mu moso bamboo carbon sequestration forest projects being carried out by Zhejiang A & F University. The 20 year project (January 2008 to December 2027) has the potential to produce 8,155 t CO_2—e net carbon credits which will traded in the Huadong Forestry Exchange (HFX). It was one of the six carbon credits suppliers in the first forestry carbon sequestration trading pilot in November 2011. The establishment of HFX has promoted the development of Zhejiang forest carbon sequestration greatly while it has worked closely with CGCF. The six carbon sequestration projects under the leadership of CGCF are traded in HFX, serving as the source of the first batch of carbon credit in China, and have successfully combined the supply and demands of carbon credit.

However, it should be noted that the net area recognised by the State Forestry Administration is only 47.7 ha^2. Because the mountain has widely distributed and fragmented stones, a considerable part of area surrounding the stone was disregarded at the time of area accounting. Notwithstanding, the actual afforestation area is nearly 86.7 ha^2—the uniform standard for afforestation operations—and the afforestation company actually paid the land rent for 86.7 ha^2. Before afforestation, the mountain was a barren shrubbery forest owned by the village collective. The afforestation company leased it at the price of US$300 per year/ha^2, the lease period would last for 40 years, and the leasing fee was paid in a lump sum in 2007. The afforestation project gets a subsidy of RMB 600,000 from CGCF; in addition, the afforestation project also received a total subsidy of RMB 105,000 in 3 years from Lin'an City Forestry Bureau. The project mainly employs a labour force from nearby foresters.

The project site Lin'an City is a relatively affluent region in eastern China, with convenient transportation and more communication between farmers and outside world markets. Farmers in Lin'an City realised the annual per capita income of US$2,321 in 2007, while the average rural farmer's per capita annual income was less than US$700 during the same period. Zhejiang is very suitable for the growth of bamboo; as a Chinese saying goes: Chinese bamboo is the best in the world, Lin'an bamboo is the best in China. Bamboo forest management is one of the important incomes for local farmers, and the Zhejiang bamboo industry achieved an output value up to US$4.2 billion in 2009.

According to different management objectives, bamboo forest operated by local farmers includes shoot-producing bamboo forest, timber forest and bamboo-shoot dual-use forest. The three kinds of bamboo forests require different management measures. Among these, shoot-producing bamboo forest has the best economic benefit. It mainly generates bamboo shoots as a food product and, in order to ensure more porous soil and better water and gas permeability, the forestland requires deep harrowing every year. In addition, shoot-producing bamboo forests need to adopt coverage technology to ensure higher shoot emergence rate and shoot emergence before the Spring Festival, because bamboo shoots can sell at the highest price before the festival. But coverage technology means the shoot-producing bamboo forest has relatively serious pests, so farmers must use pesticides to control pests and diseases. In order to promote bamboo shoot growth, farmers are generally required to use fertilisers.

Bamboo forest management based on the goal of carbon sequestration, as the business operation sets the goal of arbor stratum which has the largest fixed carbon content, soil preparation does not allow deep plowing but just strip or block soil preparation to ensure the minimum destruction of carbon pool in the soil. After bamboo plantation, it is not allowed to casually harrow land. In order to reduce carbon leakage, application of pesticides, fertilisers or mechanical operations are also prohibited. Wang and Shen (2011) summarise the difference between traditional bamboo forest management and carbon sequestration bamboo forest in management modes, as shown in Table 4.4.

Wang and Shen's study (2011) also points out that farmers risk a reduced net income from changing to a carbon sequestration bamboo forest management mode. Shoot-producing, timber and dual-use forest have the potential to reduce net income per hector by US$952, US$742 and US$909 respectively. Therefore, the carbon sequestration bamboo forest has a relatively higher opportunity cost but, if the price of forestry carbon sequestration is not high enough, farmers will have reduced revenue from operating carbon sequestration bamboo forest.

Table 4.4 Comparison of the phyllostochys pubescens management ways between carbon oriented and current management

Management model	Management objectives	Land preparation mode	Seedlings plantation	Operational tools	Day-to-day management and protection	Rotation mode
Carbon sequestration aforestation management	Ecological benefits, arbor stratum has the largest fixed carbon sequestration	Slope ≦25°, strip or block land preparation, strip width = 3X pit width; minimum damage on soil carbon	The seedlings should be planted in nearest places, growing well; 750 strains; without fertilisers	Try to use manual operation; any use of machinery may cause carbon emissions, please take detailed records	Chopper weeding, chopped weeds are left on woodland; apply organic fertilizer/ manure; dig the degraded shoots, ensure forest quality; assign specialised personnel for management and protection	Cut down 3-year old bamboo and leave 2-year old bamboo, with an annual output of bamboo 15.0–18.0 T
Current management mode	Bamboo timber, bamboo shoots have maximum net total economic benefits	Complete harrowing and deep plowing; not considering the impact on environment and soil carbon pool	The seedlings growing well, not considering carbon leakage; 300–450 strains; applying chemical fertiliser	Conditions allow general machinery operations	Mostly use herbicides; use the fertiliser of low price but high carbon emissions; natural growing; almost no specialised personnel for management and protection	Cut 4-year old bamboo and leave 3-year old bamboo, cutting every 2 years, the yield of bamboo of 900 t

Description: The data are sourced from collected information of interviews with experts in phyllostochys pubescens carbon sequestration forest and carbon sequestration forest operators

4.5.3 *Case Three: Yunnan Pu'er Forest Carbon Sinks and Forest Management*

Pu'er is a prefectural city in Yunnan with 67 % of its region covered with forests. While rich in forestry resources and biodiversity, its GDP per capita is about 6,000 Yuan, much lower than the provincial as well as the national averages. In November 2012, the CGCF Economic Promotion Centre was set up in Pu'er, Yunnan, as part of a specific action to echo the requirements of scientific development policy and sustainable development strategies. It aims to foster green growth in Pu'er and to support Pu'er to be transformed into a green economic demonstration zone. However, the carbon forest in Pu'er is currently not under CDM. Arguably, the carbon forests under CGCF are mainly aimed at improving public understanding of low carbon. Nevertheless, Pu'er can make full use of this opportunity to promote the carbon credit economic promotion centre into a not-for-profit platform to assist and educate the public to reduce emissions and put the 'low-carbon lifestyle' into practice. Pu'er will develop forest carbon sinks and sustainable forest management through this platform, to foster an ecological service market mainly on carbon credits trading and set an example to Yunnan and surrounding areas. More strict forest carbon sequestration projects under CDM should be developed in the future.

In addition to the CGCF Economic Promotion Centre, the establishment of the Southwest Monitoring Center for Carbon Inventory Methods and the launch of the National Carbon Sequestration Afforestation Pilot Project were held in Kunming, the capital city of Yunnan, on November 17, 2010, with pilots conducted in Yunnan, Shanxi, Zhejiang and other provinces. The reasons for establishing the monitoring centre in Kunming were that Yunnan has a very high forest coverage rate (over 50 % with 371 million mu of forestland in total, accounting for 62.8 % of the provincial areas and ranked as the second over the whole country), a variety of forest types, and complex afforestation species that have different CO_2 fixation quantities. Launching such a pilot project in Yunnan, especially in Pu'er, is useful for monitoring the CO_2 sequestration ability of various tree species, thus to choose the tree species which can sequester the most CO_2. CGCF has made use of the carbon sequestration forests (Pinus khasys, Betula alnoides) in Pu'er to attract public donations and encourage increasing carbon sinks as well as reducing emissions; they sent special Spring Festival greeting cards to the people who had donated, enclosing an honorary certificate for buying carbon credits.

During a research symposium about forest right mortgage loans in Pu'er, involving the People's Bank of China Pu'er sub-branch, local CCB, Agricultural Development Bank Of China (ADBC), Fudian Bank, Agriculture Bank of China (ABC), Postal Savings Bank Of China (PSBC), rural credit cooperatives and other financial institutions made introductions, the research team learned that these local financial institutions have made several achievements under the leadership of People's Bank of China Pu'er sub-branch. For example, a forest circulation market has been built up, small forest right mortgage loans and joint credit has been established in the form of Company + Farmer + Base + Credit + Insurance. Although these 3 years saw 1.729 billion yuan released, there are many difficulties. Firstly, it is somewhat difficult to manage the collaterals since the forest right is quite small, and its decentralisation made the loan officers overworked and difficult to track. In the meantime, farmers may cut down the forest privately. Secondly, it is difficult to deal with these collaterals because The Real Right Law has made rules about forest assets to be traded as collaterals. Thirdly, some collaterals have been divorced from reality, which led to profit distribution problems between farmers and companies, and this affected the sustainability of forest right credit loans.

However, by comparing the financial support on forest carbon sequestration in Pu'er with that in Hangzhou, Zhejiang, we can easily find that the financial support in Hangzhou is greater and its operation is more orderly, while Pu'er has relatively lagged behind in both aspects, especially in the section of "Assets Appraisal". Forest right mortgage loans and the circulation of the forest property right are hinged upon reliable and reasonable appraisal of forestry assets. Many representatives of local financial institutions in Pu'er who attended this symposium stated that lack of a professional appraisal agency and service is currently the biggest obstacle to effective financial support for the scale and quality of forest carbon sequestration projects in Pu'er and Yunnan in general. It is considered as necessary and advantageous for Yunnan to have its own locally-based appraisal institution, or to establish one through regional cooperation, for providing integrated forestry financial services similar to the functions and operational mode of HFX in Hangzhou.

4.5.4 *Other Findings from the Case Studies*

4.5.4.1 Support of Talents, Intelligence and Technology

In addition to the reasons above, the successful development of forest carbon sequestrations observed in Zhejiang also lies in the support of talents, intelligence and technology. State Forestry Administration (SFA) awarded Zhejiang A & F University Forestry Carbon Measurement and Monitoring Qualification, the world first carbon and networking technology laboratory was established in Zhejiang A & F University and there are sample plots in Lin'an and Anji. It was also responsible for drawing up a carbon sink measurement standard through studying moso bamboo and forest, and this standard named the Bamboo Project Carbon Sink Measurement and Monitoring Methodology was issued Doha Climate Summit. From this, we can see the strong support of Zhejiang A & F University to Zhejiang forest carbon sequestration, especially bamboo forest carbon sequestration. In addition, Lin'an has signed a cooperative agreement with Zhejiang A & F University to build a national forest carbon sink pilot site, developing Overall Planning of Lin'an Forest Carbon Sink Construction, the first example in the country.

In contrast, we find that, even though the Southwest Forestry University provides some support to the development of Yunnan forest carbon sequestration, it seems weaker compared with Zhejiang. The mode of industry-university-research cooperation can be very useful for forest carbon projects development.

4.5.4.2 Exchange of Experience at Home and Abroad

In the aspect of experience exchange, taking Lin'an bamboo forest carbon sequestration for example, a workshop called 'Sino-African Cooperation: Bamboo Forest Coping with Climate Change' at the Doha Climate Summit launched a demonstrative project on bamboo carbon afforestation methodologies in Kenya and Ethiopia, which would strengthen international cooperation. Besides, INBAR, CGCF and Zhejiang A & F University have maintained long-term cooperation.

In Yunnan Province, The Nature Conservancy (TNC) has cooperated with the Yunnan government on many aspects. TNC entered China in 1998, has established four field offices in the Three Parallel Rivers Region of Northwest Yunnan where there is the most representative biodiversity, and is carrying out field protection projects. TNC started the Tengchong Afforestation Project with the cooperation of the Yunnan Forestry Department in 2005. Although of limited scale, it has undergone the examination of United Nations Afforestation/Reforestation Clean Development Mechanism (UNFCCC, 2013) and was the first in the world to receive a Climate/ Community and Biodiversity (CCB) certificate. TNC has been motivated to protect the Yunnan area because of its unique biodiversity. From this cooperation, there is an opportunity for Yunnan to strengthen cooperation and exchange to promote its forestry carbon sequestration.

4.6 Conclusions and Recommendations

The development of carbon sequestration projects must increasingly consider the interests of farmers. According to international unwritten standards, whether a carbon sequestration project is successful not only depends on the reduced amount of carbon emissions, but also whether it can promote the development of local communities and increase farmers' income. Some scholars hope to develop some small-scale forestry carbon sequestration projects to apply to low-income communities so as to increase the income of farmers (Boyd et al., 2007). In view of the case of the implementation of forestry carbon sequestration project in China, such a program does not necessarily increase the household income of low-income communities, but may give farmers a great deal of operational risks instead. Additionality provisions on carbon sequestration projects determine that the forestland that is eligible for carrying out carbon sequestration projects is inevitably woodland of poor natural conditions or remote from the market, and its business operations are generally non-profit or even run at a loss; only such conditions can comply with additionality provisions of carbon sequestration projects. In other words, conventional commercial afforestation organisations would not choose such woodlands; without carbon sequestration projects, this type of woodland will be always abandoned and cannot be developed. Therefore, forestry carbon sequestration is not the marketisation of a forest's ecological value in the true sense. In fact, it is not to additionally increase carbon sequestration income on the basis of farmers' total forestry income, but a compensation for the loss of economic benefits of project owners for the reduction of greenhouse gas emissions. The loss occurred in advance and required feasibility studies in the design phase of the project; but the project is a loss relative to conventional commercial afforestation, so it has investment additionality. But such investment additionality is probably caused by transportation difficulties, degradation and destruction of the land, so the project owner needs to pay some costs higher than average afforestation. Thus, the net gain is certainly lower than the market profits without the support of a carbon sequestration project. If the project has very good economic benefit, the project cannot pass the assessment on investment additionality. For example, potential benefits analysis of Argentine Patagonia artificial forest shows that per hectare of forest needs investment of US$1,150, due to the remote project site and slow-growing trees; calculated by a discount rate of 10 %, the total investment return is just US$581 per hectare in 27 years, and it is impossible for conventional commercial afforestation to operate artificial forests in this region. So afforestation projects in the region can meet the assessment of investment additionality.

The development of forestry carbon sequestration projects, under the premise of economic losses, can win against loss by relying on the one chance of revenue from carbon sequestration. At present, the carbon market has large price fluctuation, so a carbon sequestration project is recognised by some investors as venture capital. The nature of venture capital is high risk and high yield. However, an afforestation company with market experience and relatively strong capital base has the capability to bear such a high operational risk, while farmers in low-income communities are relatively lacking experience in the market and have insufficient funds. Once the development project fails, farmers' families will face a devastating economic shock.

Of course, market failure can be offset by vigorous support from the government. It is practicable for farmers as owners to participate in such carbon sequestration projects because operational risks have, in fact, been assumed by the government. For the government, they have to afforest and alleviate poverty without carbon sequestration projects, so they can use financial subsidies to make up for the loss of market risks and bear the market risks for farmers.

A forestry carbon sequestration market has been established in China and is also becoming increasingly prosperous. In the process of establishing a forestry carbon sequestration trading market, forestry tenure reform plays an important role. Carbon sequestration projects require a certain scale. However, along with the establishment of the household contract responsibility system, forestland owned by the village collective is subcontracted to farmers, but each farmer household just owns small pieces of dispersed woodland. This has obstructed the implementation of forestry carbon sequestration projects in China. A new round of reform of forestry property rights system allows the circulation or centralised management on the basis of clarifying woodland property rights of farmers; forestry land use rights have become an asset of foresters, who can not only decide on trading, but also determine whether to re-centralise rights under the village collective and change property rights into co-ownership. The most important feature of this forestry property rights system reform is showing respect to the choice of farmers themselves and allowing different forestry rights form in different regions. Currently, the forestry right system tends to be diversified in China—some farmers sell forestry rights via a forestry exchange to transfer to an afforestation company, some farmers adopt multi-household union to constitute forestland management cooperatives, and some farmers re-concentrate forestland into the village collective by their combined vote, rather than decentralise to individual farmers, by adopting a collective business model with the village collective as the basic unit.

Such diversified forestry rights circulation and management modes provide a variety of choices for farmers participating in forestry carbon sequestration projects. Farmers play very important roles in such projects, because they are the providers of land and usually woodland manager and protector. Based on different circulation modes of forestland property rights, farmers play different roles in forestry carbon sequestration which also have different impacts on their incomes. Therefore, we cannot simply believe that forestry carbon sequestration projects will necessarily increase the income of local farmers. Especially for low-income community farmers participating in forestry carbon sequestration projects, their participation mode must be carefully validated and argued in order to ensure that low-income community farmers will not bear too high operational risks.

Forestry carbon sequestration projects around the world are generally divided into projects guided by private commercialisation investment and those guided by the government. Many scholars tend to be more in favour of the more in-depth marketisation in carbon finance and prefer less government intervention. However, most of carbon finance projects are characterised by high operating risks. The main consideration of private commercial investment is whether the implementation of the project can bring huge returns. Without support of a company with adequate financial strength and rich market experience, general farmers have great difficulty in bearing such high risks, and it is even more difficult for farmers in low-income communities (see Grieg-Gran, Porras, & Wunder, 2005). Forestry carbon sequestration projects guided by government mainly consider the benefits from environmental recovery, soil conservation and community poverty alleviation. Therefore, carbon finance projects have no operational risks for government but a beneficial supplement for environmental protection and poverty alleviation finance instead (see Smith & Scherr, 2003). Many forestry carbon sequestration projects in China can improve farmers' income, mostly through the active participation of the government and its support via the technology, policy and funding and other aspects, while sharing most of the business operation risks.

Based on this fact, in order to enable forestry carbon sequestration projects to improve simultaneously the welfare of the community, reduce greenhouse gas emissions and increase farmers' income, it is argued that the government should be more actively involved in forestry carbon sequestration projects and employ its financial strength to bear part of the project operational risks.

References

Benitez, P., McCallum, I., & Obersteiner, M. (2004). *Global supply for carbon sequestration: Identifying least-cost afforestation sites under country risk considerations* [R]. Laxenburg: International Institute for Applied Systems Analysis (IIASA) IR-04-022.

Boyd, E., Gutierrez, M., & Chang, M. (2007). Small-scale forest carbon projects: Adapting CDM to low-income communities. *Global Environmental Change, 17*(2), 250–259. http://www.sciencedirect.com/science/article/pii/S0959378006000811

Cangwu County Annals Compilation Committee. (1996). *Cangwu County annals* (pp. 216). Nanning, China: Guangxi People's Publishing House.

China State Forestry Administration. (2009). *The CGCF guide for carbon sequestration project measurement and monitoring* [M] (2nd ed., pp. 69). Beijing, China: China Forestry Publishing.

China State Forestry Administration. (2012). *Points for forestry "Twelfth Five Year" plan on climate change.* http://www.gov.cn/zwgk/2012-01/20/content_2049496.htm. Accessed 08 Sept 2013.

Forestry Bureau of Cuangxi Zhuang Autonomous Region. (2007). *The world's First CDM Forestry Carbon Sequestration Project in Guangxi.* http://www.forestry.gov.cn/portal/main/s/72/content-370966.html. Accessed 12 Sept 2013.

Grieg-Gran, M., Porras, I., & Wunder, S. (2005). How can market mechanisms for forest environmental services help the poor? Preliminary lessons from Latin America. *World Development, 33*(9), 1511–1527. http://www.sciencedirect.com/science/article/pii/S0305750X05000999

Huanjiang Autonomous County Annals Compilation Committee. (2001). *Huanjiang Autonomous County annals* (pp. 323). Nanning, China: Guangxi People's Publishing House.

Li, N., Gong, Y., & Zhang, S. (2006). Forestry carbon sequestration projects in China: A synergized strategy of mitigation, adaptation and sustainable development. *World Forest, 19*(3), 1–5.

Li, N., Yang, Y., & Chen, X. (2010). The development of carbon sequestration forestry to respond to climate change – The practice and management of carbon sequestration forestry in China. *Chinese Science of Soil and Water Conservation, 2*, 13–16.

Li, N., & Yuan, J. (2011, January). Chinese forestry construction under climatic change background. *Protection Forest Science and Technology, 1*(Sum No. 100), 4–7.

Miao, G., & West, R. A. (2004). Chinese collective forestlands: Contributions and constraints. *International Forestry Review, 6*(3–4), 282–298. http://www.bioone.org/doi/abs/10.1505/ifor.6.3.282.59971?journalCode=ifre

Smith, J., & Scherr S. J. (2003). Capturing the value of forest carbon for local livelihoods. *World Development, 31*(12), 2143–2160. http://www.sciencedirect.com/science/article/pii/S0305750X03001694#

Thomas, S., Dargusch, P., Harrison, S., & Herbohn, J. (2010). Why are there so few afforestation and reforestation clean development mechanism projects? *Land Use Policy, 27*(3), 880–887. http://www.sciencedirect.com/science/article/pii/S026483770900204X

UNFCCC. (2006). Facilitating reforestation for Guangxi water-shed management in Pearl River Basin: Project design document submitted to United Nations Framework Convention on Climate Change. http://cdm.unfccc.int/UserManage-ment/FileStorage/H5218OI0ZWU4CTWLPLKEIETBIODYED

UNFCCC. (2013). *Afforestation and reforestation projects under the clean development mechanism: A reference manual.* http://unfccc.int/resource/docs/publications/cdm_afforestation_bro_web.pdf

Wang, X., & Shen, Y. (2011). Analysis of bamboo carbon sequestration afforestation management mode and its benefit. *Journal of Zhejiang Agricultural University, 28*(6), 943–948.

Yang, P. (2008). Discussion on the circulation of collective forest property right in Fujian province. *China Forestry Economics, 3*, 32–36.

Zhang, Y. (2001). Institutions in forest management: Special reference to China. *World Forests, Society and Environment, 3*, 353–364.

Zhang, X., & Wu, S. (2005). *Theory and practice of forestry carbon sequestration projects* (pp. 375). Beijing, China: China Forestry Publishing.

Chapter 5
Alternative Energy Development in Rural Chinese Communities

5.1 Introduction

Energy is the driving force for socioeconomic activities and human development; the energy sector drives economic activities, especially those which sustain domestic productivity, national progress and global competitiveness. Without energy, all development would come to a halt. While rural electrification carries an important weight in rural development, rural development further correlates significantly with regional growth and national securities—in particular, food security, social stability and environmental ecology. The complex issues relating to energy and poverty further magnify the focus on alternative energy development in rural areas; that is, credibility in meeting energy demand, sound livelihood improvement and a subsequent sustainable growth through the test of time.

Energy poverty—the energy dimension of poverty—is the unfortunate consequence of energy underdevelopment and has become a critical issue for humanity. Energy poverty may be defined as "the absence of sufficient choice in accessing adequate, affordable, reliable, quality, safe and environmentally benign energy services to support economic and human development" (Reddy et al., 2000, p. 44).

Worldwide, the speed of population growth vastly surpasses the provision of modern energy access, which exacerbates energy poverty, especially in rural and remote areas. The International Energy Agency (IEA) (2011) estimated that over 1.3 billion people live without electricity access and another 1 billion people encounter sporadic or unreliable electricity supply, whereby 2.7 billion people, lacking modern energy services, rely solely on traditional biomass for daily cooking and heating (Table 5.1).

The lack of modern energy and services inhibits people's basic livelihoods and household income-generating potential. Additionally, it hampers other social provisions such as sanitation, health care, and education, and impacts upon general pollution levels and deforestation. Currently 1.3 billion people subsist on under

Y. Zhu et al., *Transforming Rural Communities in China and Beyond*,
DOI 10.1007/978-3-319-11319-7_5

Table 5.1 World electricity access in 2009—regional aggregates

	Population without electricity (million)	Electrification rate (%)	Urban areas with electrification rate (%)	Rural areas with electrification rate (%)
Africa	587	41.8	68.8	25.0
North Africa	2	99.0	99.6	98.4
Sub-Saharan Africa	585	30.5	59.9	14.2
Developing Asia	675	81.0	94.0	73.2
China and East Asia	182	90.8	96.4	86.4
South Asia	493	68.5	89.5	59.9
Latin America	31	93.2	98.8	73.6
Middle East	21	89.0	98.5	71.8
Developing countries	1,314	74.7	90.6	63.2
World[a]	1,317	80.5	93.7	68.0

Source: IEA (2011, p. 473)
[a]The world total includes OECD and Eastern Europe/Eurasia

US$1.00 per day; the majority of the impoverished socioeconomic class cluster in and around developing Asia, Sub-Sahara and Africa. Furthermore, 70 % among the disadvantaged are women who are forced to deplete the environment (e.g. cut trees for fuel) due to inadequate consumption patterns for their daily subsistence.

For the next 20–30 years, many rural and urban poor in the developing world will likely continue to depend on traditional biomass. Therefore, it is clear that energy strategies should be well designed and implemented within the framework of national policies and local coherence to sustain a sound energy transition and holistic development path (Yumkella, Bazilian, & Gielen, 2010). The Busan HLF4 Declaration (2011) calls for development effectiveness whereby "development is driven by strong sustainable and inclusive growth". While development strategies should not focus singly on economic outcomes, Simon (1997, p. 185) stated that sustainable development needs to embrace a sound human dimension which is:

> …the process of enhancing individual and collective quality of life, that satisfies basic needs, is environmentally, socially and economically sustainable and is empowering in the sense that the people concerned have a substantial degree of control over the process through access to the means of accumulating social power.

Behind every challenge and inequality, there lie in wait opportunities for improvements. The current peak oil prices and climate change situation could prompt accountable alternative energy strategies, steering demand away from fossil fuel dependency. Rather than being a set of competing demands on nature, a successful energy strategy aims to harmonise social development within nature's ecosystems, avoiding resource constraints, achieving more with less, for the benefits of local well-being,.

In China's 12th Five Year Plan, the importance of reducing poverty and improving environmental management is addressed by the relative indicator system, underpinned by the emphasis on effective ecological development and comprehensive capacity building. Better social management is highlighted by constructing greater social harmony such as the improvement of labor productivity and general public services for both urban and rural residents. A more reasonable income distribution system needs to be implemented quickly—in the rural context, the per capita disposable income of rural residents needs to rise by an annual average of over 7 % in real terms.

With specific regard to clean energy in the rural environment, non-fossil fuel is to step up to 11.4 % of national primary energy consumption, whereby energy consumption per unit of GDP is to be cut by 16 %, CO_2 emission cut by 17 % and forest coverage rate to rise to 21.66 % (forest stock to increase by 600 million cubic metres). At the local level, the rural development target is to offer sustainable rural energy production and service systems that contribute to poverty alleviation with practical income-generating opportunities. Cost efficiency and resource productivity can be explored through life cycle approaches involving materials and service exchange between local actors, while the challenge remains on what and how to promote the increased integration and exchange of energy, material flows, and the involving human factors. To identify the implications on the ground, the research is designed to address three key research questions:

1. What resources and measures should be taken to break the vicious cycle of the poverty trap and ecosystem degradation in energy development?
2. What clean energy technology or service innovation is affordable and scalable and can also improve local livelihoods with a focus on entrepreneurship?
3. What conditions can be brought about to encourage community development and green entrepreneurship in rural areas, especially for the poor?

In order to explore the key issues, this chapter is structured as follows; we first consider the wider view of the world energy backdrop to provide an understanding of core energy issues and actions to be taken within China's system. The next section reviews China's energy strategy and rural development situation, with a further exploration of renewable energy potential for livelihood improvement. Then we present a literature review of the relation between energy and poverty, while the alternative 'whole system approach' in rural energy development is further discussed. New business approaches are then discussed, that unleash community assets, collective strength and renewal energymass potential, with a focus on small hydropower (SH). The last section reflects on the findings and challenges, while a final system review is provided to deepen the framework of energy poverty alleviation and to highlight areas for future research.

5.2 World Energy Backdrop

The McKinsey report (2011) stated that the world could experience a considerably longer period of economic downturn with a plunge of global GDP of $1.5 trillion if oil prices stay high for several years, topping other factors such as climate change, financial crisis and political instability. Today, the timing of global economic recovery remains uncertain. Although developing economies are experiencing economic recovery, the growth momentum relies heavily on the unsustainable consumption of the northern (OECD) countries. Many developing countries still face critical exposures in the energy dimension, such as high vulnerability towards the international energy pricing and trade movements, energy inefficiencies, infrastructures and structural issues within the countries' whole energy systems.

In the global efforts of tackling climate change, the IEA (2011) marked out global investment environment that, by the 450 Scenario,[1] four-fifths of the total energy-related CO_2 emissions permissible until 2035 are already 'locked-in' by our existing capital stock and infrastructures (i.e., power plants, buildings, factories, etc.). It further explained that by contrast the energy-related infrastructure in place today will generate all the CO_2 emissions allowed, hence all future investments and expenditures must be all carbon neutral, meaning extremely high opportunity cost is to be envisaged, should interventions fail to be delivered by 2017 according to the 450 Scenario per se.

Instead of current policy, the alternative New Policy[2] Scenario aims to bring down the long-term average temperature increase to 3.5 °C or more. Should the governments and the societies at large fail to act upon New Policy, such as restraining capital investment flowing into harmful, energy-intensive industrial sectors and power plants, or encouraging low carbon investments in value-innovation to conserve our natural capital, an additional cost of US$4.3 will therefore be incurred in order to compensate for the increased emissions after 2020, for every US$1 investment avoided.

Should governments fail to implement new policies beyond those already planned till 2030, the IEA (2011) projected in another study that the world's total energy consumption by 2030 will increase by over half (53 %), CO_2 emissions will increase by 55 %, with a resulting energy mix dominated by fossil fuels (80 % shares), while large populations (1.5 billion) of the world's poor will persistently lack access to electricity and another 2.5 billion will lack basic modern energy services for heating and cooking.

[1] "450 Scenario: A scenario instated in the World Energy Outlook, which sets out an energy pathway consistent with the goal of limiting the global increase in temperature to 2 °C by limiting concentration of greenhouse gases in the atmosphere to around 450 parts per million of CO_2" (http://www.iea.org/publications/scenariosandprojections/).

[2] "New Policies Scenario: A scenario which assumes broad policy commitments and plans announced by countries, including national pledges to reduce GHG emissions, plans to phase out fossil-energy subsidies, and other measures to implement these commitments which are yet to be identified" (http://www.iea.org/publications/scenariosandprojections/).

An emerging middle class (about three billion people, primarily from Asia) will drive energy consumption in the future. Developing (non-OECD) countries, particularly China and India, will lead economic growth but consequently will also lead energy and other resource consumption (e.g., materials, food, water) (IEA, 2011). The projection of world total energy consumption of non-OECD emerging economies will account for 62.6 % of the world's total by 2035 with an average annual increasing rate of 2.3 % compared to the rate of OECD countries' at 0.6 % over the projection period (IEO, 2011).

A truly global and integrated energy technology transformation is essential to address the cross-cutting challenges of climate change, poverty and energy security while at the same time meeting the growing energy needs of the developing world as well as economy recovery (IEA, 2011). That being said, the low carbon energy technology revolution and 'decoupling' must be readily implemented to curb the rebound effect—where increased efficiency leads to lower costs, which drives further consumption.

Instead of reiterating the theories of technological evolvement, von Weizsaecker, Hargroves, Smith, Desha, Stasinopoulos (2010, p. 3) envision a practical picture of "the whole systems technologies, infrastructures, legal rules, education and cultural habits interacting to produce economic progress while conserving a healthy environment". The whole system approach can also overcome the rebound effect by bringing about efficiency and a lifestyle approach of sufficiency as the new mantras for quality of life, with the combination of socially-benign policy framework and economic instruments to embrace a holistic, socially inclusive growth.

5.3 China's Energy Situation

Given its globally dominant position in both energy consumption[3] and coal power generation, China's coal consumption will surpass the rest of the world's total by the year 2017 (IEA, 2012). Facing the pressure of true value pricing systems on energy and high carbon, China must undertake major steps such as investing largely in cleaner coal (i.e., carbon capture and sequestration) and renewable technologies with a view to reinforce its energy efficiency in both power generation, industrial utilisation and building sectors, along with networks and infrastructure improvement.

To comply with climate change governmental responsibilities, China is giving priority to low carbon development that firstly measures energy productivity, targeting its indigenous resource productivity and energy efficiency by restructuring

[3] China's energy consumption in 2011 accounted for 20.3 % of the world's total (IEA, 2011), from the composition of its domestic primary energy production; coal power took up 81.3 %, crude oil 10.4 %, natural gas 4.5 %, hydro 3.2 %, nuclear and others 0.3 % (China National Bureau of Statistics, Department of Energy Statistics, 2011).

both primary and secondary energy intensive (also polluting) sectors such as iron, steel, cement, chemicals, building and construction. The country also aims to fasten rural energy and infrastructure development parallel to the encouragement policy in agricultural and land productivity management to upscale rural livelihoods.

Among others, renewable energy and corresponding technology developments receive great impetus from the current 12th Five-year Plan, which entails the need to strengthen resource conservation, adequate supplies, and resource allocation based on ecological development and poverty alleviation. Notably, the expansion of domestic consumption is regarded instrumental for economy restructure, which is to drive the national economy away from overt export dependency. We know that inclusive growth patterns require substantial efforts to prevent rebound effects by decoupling growth from environmental-social impacts.

Thus, country systems should be designed and implemented to ensure sustainable interim production and use of traditional energy, with special focus on smooth social transition, that is, sound rural development that shoulders the objectives of poverty alleviation and the prevention of a consumption-led growth trajectory, to sustain the country's long-term economy. After all, the costs of providing assistance to pro-poor and alternative rural energy development may turn out far less than those of coping with social unrest, instability and migratory urban sprawl associated with rural poverty (Birol, 2007).

5.4 Renewable Energy

Recognising that the lack of affordable and efficient rural energy services hinders other provisions such as sanitation, health, education and the supporting infrastructures on which income-generating activities depend, China started the first round of rural electrification in 1998. In 2002, the National Development Reform Commission (NDRC) further took the lead in China Township Electrification Program (Song Dian Dao Xiang) with the target to provide universal electricity access for 1,065 townships spread over 12 provinces; the program continued until 2010 and has benefited 23 million rural populations.

The objective of the Song Dian Dao Xiang is to stimulate rural economic activities through productive use of electricity with a special focus on the decentralised (grid improvement) electrification for remote rural areas. The program adopted both on-grid extension and off-grid stand-alone systems with renewable technologies inclusive of wind, photo voltaics (PV), PV/wind combined, hydropower (H) and SH designed according to local conditions. Remarkably, the fuel-switch programs (stipulated in 2003) from community SH were widely established: electricity transmission was connected to the grids and households for cooking and heating to displace burning solid fuels. Special subsidies were also given to households that could not afford the purchase of basic electrical appliances.

China lends its hydropower expertise and fosters cooperation worldwide. Zhejiang Province[4] is one of the most populated provinces because of its established SH, and the Jingning County is celebrated as the 'Hometown of Chinese rural hydropower'. Overall in China, hydropower (annual power generates about 130 billion kWh) has become the important constituent of the national power supply, supplying nearly half of the country's land area and a quarter of the nation's population. Globally, renewable energies are becoming the centre of attention; the share of non-hydro renewable energy used in power generation is to expand from 3 % in 2009 to 15 % in 2035 mainly driven by China and the European Union. Most renewable energy programs are subsidised with annual renewable electricity subsidies increasing to nearly US$180 billion by 2035 (IEA, 2011, p. 532) in order to compete in electricity markets. While energy efficiency and renewable energy are said to be the twin pillars of sustainable energy policy, renewable energies are expected to bring long lasting benefits of smooth social transitions in terms of energy security, environmental protection and improved rural livelihoods (Wang, Jing, Zhang, & Zhao, 2009).

Since 2005, a number of laws and regulations have been promulgated and updated for energy conservation and emission reduction such as 'Renewable Energy Law', 'Energy Conservation Law' and 'Medium- and Long-Term Development Plan of Renewable Energy'. In the meantime, China has successfully implemented national policy frameworks and corresponding market mechanisms (i.e., domestic carbon trade, subsidies and feed-in-tariffs [FIT]) for wind, biomass and PV to boost industrial energy efficiency and renewable power generation since renewable energy industries had been elevated for national competence and as the country's economic growth engines.[5]

Continuous efforts in energy policies and financial mechanisms intended to deepen rural energy development are picking up the pace, such as the most recent forest carbon schemes (2011) and the improved biomass FIT with the aim of regulating the market and increasing resource productivity in agriculture and farm forestry (an average increase of RMB 0.07/kWh, NDRC price [2010] No.1579),[6] while

---

[4] SH in China defines installed capacity $\leq$ 50 MW with an extension hook on local grid networks to balance additional power generated or emergency power supply. The salient benefits include off-take tariff and low construction costs with an investment return usually less than 10 years. In Zhejiang, from the period of 1994 to 2002, a total of US$1.33 billion was invested for 1,058 MW, more than 70 % shareholdings from private enterprises. The benefits include scalability, cost effectiveness (ROI 10 to 15 % p.a.), the off-take tariff, cheaper construction costs, longer service life and operating hours (i.e., 3,000–4,000 h annually in Zhejiang Province) (Zhao & Zhu, 2004, ICSHP).

[5&6] Source: Complication of China policies and Regulations on New Energy And renewable Energy (1986–2011).

[6] Golden Sun PV subsidy program ran through from 2000 to 2011 on a project-by-project approval base. The objective of the program is to boost PV industry through fiscal subsidies, scientific and technological support and market incentives. The program offers 70 % capital subsidies for off-grid installations in remote areas and 50 % capital subsidies for grid-connected installed capacity >300 kWh.

the 'Golden Sun' program was stipulated to promote both on-grid large-scale and off-grid decentralised installations.

5.5 Untapped Economic Potential—The Biomass

In the national science and technology development planning (National Platform for medium- and long-term science and technology development planning, 2006–2020), strong emphasis was laid on the huge economic potential to be brought about by biomass. As a rule of thumb, the best profit (or saving) is to utilise resources that we usually waste or don't use. The position of biomass power generation is rated second to hydropower among the renewable energies in China. The potential biomass energy in China alone in 2004 amounted to 3,511 Mtce[7] and 460 Mtce from accessible biomass (Shen, Liu, Yao, Liu, & Lucas, 2010). The biomass energy resources in China mainly come from by-products, waste and residues from agriculture, forestry (i.e., straws, husks, residues, manure, energy crops, sewage and organic waste).

Taking crop straw for instance, the total amount potential is 600 million tonnes/year, of which half of the quantity goes to paper (packaging) industries and the remaining half (300 million tons) is suitable for energy production, which is about 150 million tonnes/year of coal equivalent. The potential forest residues for energy can reach up to 300 million tonnes/year (about 200 million tons of coal equivalents).[8] Overall in China, it is estimated that the annual quantity of biomass potential amounts to 500 million tonnes/year and another 580 million tonnes/year of agricultural manure and agro-industrial organic waste (including waste water). As for forestry energy, it is projected to reach 1 billion tonnes through widened afforestation efforts[9] (von Weizsäcker et al., 2009).

Enormous social-ecological benefits are expected to result from a quantum leap, as announced in China's science and technology development plan (2006–2020), whereby during this period biomass-based rural electrification shall benefit ten million rural residents without electricity access and generate additional rural income at about RMB 100 billion Yuan. In the meantime, forestry revenue shall increase by RMB 50 billion Yuan per annum through continuous afforestation and eco-service systems, while two million jobs are to be created in the field of corresponding renewable energy sectors (Fig. 5.1).

5.6 Energy, Poverty and Development

As stated by Barnes and Floor (1996), the first energy-related economic issue is the extensive reliance and inefficient production of traditional energy sources, such as the direct burning of fuel wood and agro-residues that causes economical,

[7] Mtce is the abbreviation of million tons of coal equivalents and 1 Mtce is equal to10^6 tce.

[8 & 9] Source: Complication of China Policies and Regulations on New Energy and renewable Energy (1986–2011).

	Municipal Waste*	Industrial Waste	Primary Solidfuel**	Biogases	Liquid Biofuels	Geothermal	Solar Thermal	Hydro (LHP & SH)	Solar PV	Tide, Wave, Ocean	Wind
Gross Electricity Generation (unit: GWh)	0	0	2351	0	0	153	0	615640	321	0	26900
Gross Heat Production (unit: TJ)	0	0	11869	0	0	0	0				
Unit	*TJ*	*TJ*	*TJ*	*TJ*	*1000 (tons)*	*TJ*	*TJ*				
Production	0	0	8152659	322384	1868	75191	300633				
Domestic Supply	0	0	8152659	322384	1868	75191	300633				
Transformation loss	0	0	53626	0	0	5508	0				
Electricity Plants	0	0	33845	0	0	5508	0				
CHP Plants	0	0	0	0	0	0	0				
Heat Plants	0	0	19781	0	0	0	0				
Energy Industry's Own Use	0	0	0	0	0	0	0				
Industry	0	0	0	0	0	2950	0				
Transport	0	0	0	0	1868	0	0				
Residential	0	0	8099033	322384	0	37058	0				
Commercial and Public Services	0	0	0	0	0	20170	0				
Agriculture / Forestry	0	0	0	0	0	9003	0				

* Municipal Waste: the split for renewable & non-renewable is available on IEA website

** Primary Solidfuels includes Charcoal

Fig. 5.1 Renewable and waste production (except biogas) in P.R. China, 2009 (Source: http://www.IEA.org)

environmental and health threats which currently exist in mass scale in our society. The second issue is the widening gap between energy demand and supply thanks to a burgeoning world population, which causes extreme shortages and disruption of modern energy supplies (i.e,. electricity, liquid fuel or gas), thus affecting quality of life aside from the issues of security, equity and social inclusion. Birol (2007) highlighted the resulting energy poverty.

Wang and Hu (2010) concluded that most rural households use traditional solid fuel energy, while the energy consumption per capita is in positive correlation to income per capita. The rural poor are often forced to resort to environmental depletion due to inadequate clean consumption patterns. Research indicates there is a strong relation between rural household energy consumption and the use of commercial energy at different income levels (Jiang & O'Neill, 2004; Lu & Li, 2006). In rural areas, the household consumption of coal and traditional biomass declines in contrast to the increasing usage of gas and LPG, whereby the demand for electricity shows little change in relation to the upward trend of household income. At the same time, coal, diesel and electricity remain as strong substitutes for traditional biomass among rural mechanisation and income-generating activities.

It is clear that access to energy is not just about quantity but more essential are the soft systems, calling for the quality, scalability as well as the infrastructure to support the corresponding energy products and services. The World Bank claims that large-scaled rural electrification can induce industrial growth in low income rural areas, while the feasibility of a higher economic rate of return for off-grid installation or grid-extension is 'far from clear' (World Bank, 2012). Evidence as seen in China shows that most rural economies stagnate and largely lag behind those in urban areas, even with grid-electricity access. Furthermore, Kaygusuz (2012) explained that "rural electrification has not, by itself, triggered industrial growth or regional development"; he further concluded that "rural electrification is economically justified only when the emerging uses of electricity are strong enough to ensure sufficient increase in demand to produce a reasonable economic rate of return on the investment" (i.e., the upfront cost to purchase matching household appliances or productive assets).

More than 30 million people without electricity access benefitted from China's universal energy access[9] programs ('Brightness Program', 'China Township Electrification Program', 'Rural Grids Restructure', 'Rural Water and Electrification Program') since the late 1990s with on-going efforts until now. The government continues its efforts in upgrading rural energy structures and energy consumption patterns for the 727.5 million (by 2007) rural populations whose livelihood depends heavily on burning straw and traditional biomass due to the constraints of technology and affordability, lack of supporting infrastructure and services (Luo & Zhang, 2008). That being said, the challenge remains in transforming China's rural energy consumption patterns while improving people's livelihoods at the same time. The approach to this challenge is to tap into the unused and abundant biomass, given that the biggest saving (earning) would come from 'utilising stuff that people don't use' with renewable energy production and the related business service potential, to unwind the carbon (energy) intensive growth trajectory.

Balancing common-pool resources, private with public goods means to a large extent balancing economic aspirations with ecological imperatives. We then ask ourselves to what extent are energy structural changes and the output growth trend related? How do the influential factors such as income, consumption and social learning interact with energy transition? On the subject of sustainable resource management, von Weizsäcker has mentioned: "Sustainable development has a chance if we learn to extract five times more for human development wealth from one hectare of a footprint or from one unit of natural resources, with a whole system approach to increase resource productivity and develop a mechanism of fair distribution" .[10] An integrative whole system approach, as introduced in Chapters 3 and 4, is necessary to propel sustainable growth.

---

[9] Universal energy access covers basic services but not productive uses for income-generating activities (basic service includes cooking, heating, lighting, and other household electric appliances such as TV and radio sets.).

[10] From an open seminar by Prof. Von Weizsäcker in March, 2009, Tokyo, about resource management and the 3Rs for a sustainable Asia with a focus of decoupling consumption from growth in the context of sustainable resource management.

5.7 Alternative Energy Development with the 'Whole System Approach'

Alternative energy development is about the effort of providing sustainable energy services with decentralised service systems, whereby the primary energy production is generated from local renewable resources and waste streams. This section examines tangible energy production and service systems, in the Chinese rural context, with a 'whole-system-approach', as described by Ness and Vogiatis (2013), which can offer affordability, synergistic livelihood options, pollution prevention and climate change mitigation in tandem with the Chinese government's goal (12th Five-year Plan, low carbon development) to develop a resource efficient and environmently friendly (REEF) society alongside the theme of the monitoring systems (REPI) to monitor the development progress.[11]

Von Bertalanffy is one of the pioneers of 'general systems theory', which promotes the notion of holism rather than over-specialisation, in keeping with the 'Aristotelian' world view that the whole is more than the sum of its parts; he emphasized that "order or organization of a whole or system, transcending its parts when these are considered in isolation…is a fact of observation encountered whenever we look at a living organism, a social group, or even an atom" (Von Bertalanffy, 1972, p. 408).

Recognising the interactions surrounding energy, society and environment in the modern world today, the provision of adequate energy services is seen as a precondition for socioeconomic growth and human development. Within the sphere of product and service systems, energy services enable the transformation and embodiment of other resources such as capital, materials, skills, and technology innovation on which the social construct depends. Energy services also result from the bundle of various technologies, infrastructures, capital, materials and energy forms alike. As described earlier, worldwide energy systems are facing significant challenges, such as the need to increase cost effectiveness, energy productivity, decoupling the energy consumption trend from growth whilst meeting greenhouse gas (GHG) emission targets. The challenges call for effective energy service systems (Fig. 5.2), which requires innovations, tools, technologies, and policies in combination with qualified human capital to capture the dynamics and networks between energy resources, demand, consumer behaviours and, in parallel, a sound adaptive capacity from the system users (product and service users) to derive value from the systems.

[11] REPI (Resource and Environment Performance Index) was developed by the Chinese Academy of Science following the principle of eco-efficiency (see UNESCAP, 2010). The 'REPI' indicator system measures regions' (or the country's) level of resource use, by quantifying the intensity of resource consumption (or pollutant discharge performance of resource) in relation to GDP output).

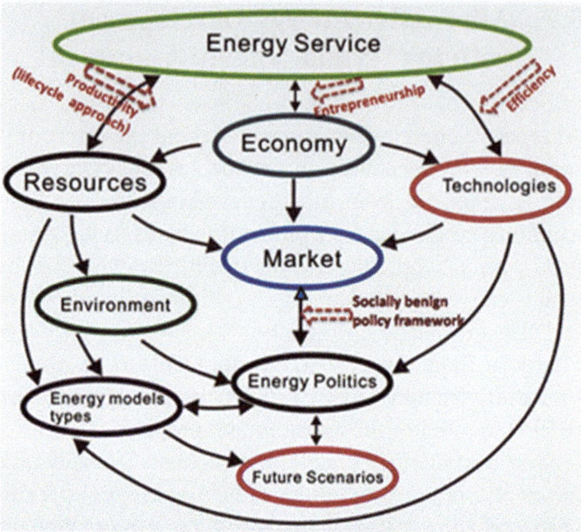

Fig. 5.2 Energy service and the system (Courtesy Energy Economics Group, Vienna University of Technology)

Yumkella et al. (2010) emphasised that the prime objective of the sustainable energy systems is to improve the long-term performance of a nation's energy sectors, addressing climate change and energy security concerns especially in developing economies (sustainable energy for all).

With a rural setting in mind, particularly in remote or low income areas, the energy system should involve a transdisciplinary approach in aspects such as engineering, economic, ecological, social and business management beyond the renewable and local context (as was adopted by our research team). The design and implementation can embrace alternative human factors (the human dimension) within the energy development frameworks which eventually foster income creation or entrepreneurship opportunities where the poor would be empowered through their increased capability to 'own', to 'make accountable decisions', and to 'take control of their own future'. Thus, their self-esteem is elevated with a sense of social inclusion through the empowerment processes. Parallel with energy and rural development, China's 12th Five Year Plan also emphasises the State's efforts to alleviate poverty through development with great efforts into water conservancy, agricultural productivity and rural infrastructure, whereby governments at all levels strive to accomplish practical results for the people (NDRC, 2011).

Improved energy services result in increased household and industrial productivity, better living and health conditions, especially for women and children, thanks to better air quality from improved cooking and heating conditions. Most prominently, such improved rural energy services induce the uplifted community well-being and economy (Cecelski & Unit, 2000) in common with improved social learning such as education and community knowledge acquisition, which further generate a greater ability to pay for productive assets and improved energy services to run household work, agro-processing or micro ecotourism. As Warr and Ayres (2006) have demonstrated, long-term economic growth and output productivity evolve alongside significant energy transition (structural changes) in response to the positive switch of production factors generated by innovations and externalities (external shocks). The effect of output is largest with respect to energy service (useful work) and smallest for labour, which indicates a productivity shift from labour intensity towards resource productivity prompted by technology or business models and supply chain innovations.

Today we are facing extreme externalities on climate change combining the endogenous humanity issues arising from socioeconomic and environmental problems. Such urgency, however, presents extraordinary opportunities for all societies to break through old norms and seek solutions for more equitable reallocation of resources and public goods with energy efficiency complemented by appropriate production consumption technologies and service systems.

5.8 Energy Productivity and the Whole System Approach

Whole system design with integrative approach reaps cost effective benefits for all operating systems; whether it is for building, business, agriculture or production processes etc., an integrated all-round approach is essential to avoid waste and excess along the system networks (von Weizsäcker et al., 2009). In the pro-poor energy development setting, a thorough 'demand analysis' (Pachauri & Spreng, 2004) can enable 'need-based' design constructs tailored to the needs of the poor, with prioritisation and ranking (Cecelski & Unit, 2000) according to situations, time variation or locations that enhance competent productivity for the poor, leveraged with cost appropriation thus affordability.

Energy productivity is a term that measures the output and quality of goods and services per unit of energy input (material input per service unit), it comes from reducing the energy amount required for input production or the increased output quality and yield. Von Weizsäcker et al. (2009) claimed that a multifold increase of productivity in agriculture, up to factor 20, could be achieved with a whole system approach. Furthermore, it was asserted that conventional farming can achieve a factor five improvement in energy productivity through a combination of direct and indirect initiatives and investment in renewable energy options (Tables 5.2 and 5.3).

5.9 The Role of Business and Social Enterprises

Social entrepreneurship and enterprise-led development have emerged as innovative and crucial instruments for the effectiveness of pro-poor growth. Enlightened thinkers have suggested 'enterprise solutions to poverty' (Harvard CSR initiative, 2007) or 'profiting with the poor' with a technology driven process (Kandachar & Halme, 2008) and a consideration of the role of the private sector in large-scale market development in the 'bottom-of-the-poor' market. Gradl and Jenkins (2011) further

Table 5.2 Village case profile and general model

Villages	Mangjing Village (*Blang* minority), Lancang, Yunnan	Daxi Village (*She* minority) Jingning, Zhejiang	Baisha Village Linan, Zhejiang
Major characteristics of community business, products & services	One village, one Pu'er tea product	H & SH stations	Local tourism & farm forestry
	From tea plantation to sales distribution	Farming	Diversified multiple income streams
	small cooperatives & family-run businesses, numerous brands	Tourism in plan	Vigorous small & medium businesses. Environment & enterprising spirit
	'Brilliant Group' – local leading enterprise major share of local tea production and employment		
Transitional local economies and livelihoods	Developing Commercialisation & Branding	Low income level – conventional farming	Industrial specialisation
	Tea forest biodiversity in process	High income level – SH, large-scale farming & related sectors	Land management in place
	Tourism entrepreneurship in early stage		Branding & channel partnership (tourism) in plan
Community Assets Community Strength Community Capacity	Strong ethnic culture	Strong SH technologies & investment expertise	Sufficient job markets
	Eagerness for social inclusion	Beautiful cultural legacy inbuilt	Strong female voices present
	Strong women voluntary association	Breath-taking scenery, oxygen and eon rich mountain trails	Vigorous entrepreneurialism
	Active female voices present	Low and high mountains rich with biodiversity	Organised community training program present
	Community training program in place	Water, sanitation, community centre in place	Strategic alliance with NGOs & research institutes in place
	Transport, schools, warehouse, water, sanitation		Sufficient non-farming income
			Eco forestry & biodiversity in place
			Transport, communications, schools, health centres, warehouse, water, sanitation in place

Phenomena Possible capacity gaps and challenges	Most migrant workers return to village shortly after seeking jobs outside	Substantial income gap present	Randomly developed local tourism
	No manufacturing process specialisation. No process quality standardisation	Disrupted ethnic culture largely caused by dislocation (due to universal energy access)	Energy intensive local infrastructure
	No process pooling, no resource sharing	Disruption of biodiversity & local tourism (waterfall scenery spot) from bigger sized hydropower establishments	Built environment in disharmony with nature
	Weak job market & human capital: Young generations stay in village for jobs	Unpaid female household work is common	
Local renewable resources potential	Biomass, forestry residues, residential, organic waste, tourism waste stream	Hydro, biomass, forestry residues, residential organic waste	Biomass, forestry residues, residential organic waste, tourism waste stream
Current energy source Current energy service (for households)	Fuel wood & residues	Fuel wood & residues	Fuel wood & residues
	Most households cook with traditional biomass stove	Most households cook with traditional biomass stove	Grid electricity
	Inconsistent grid electricity supply	Electricity (hydro, grid-extension)	
Current energy source Current energy service (for local business, workshop, production)	Coal, charcoal, LPG, diesel generators	Fuel wood, solid biomass, coal, LPG, grid electricity.	Fuel wood & residues
	Inconsistent grid electricity-supply		Grid electricity
Demand on energy services	Thermal energy for: food and agricultural product processing (i.e., cooking, baking, drying, heating, smoking, etc.)		
	Electricity for: agricultural product storage, refrigeration, cooling, heating, lighting, ICT, mobile, entertainment		
	Motive transportation power for: farm, family, cooperatives, small-scale industries, micro-tourism		
	Field mechanisation power for: ploughing, tilling, irrigation, thrashing, harvesting		
	Local industry and workshop mechanisation: manufacturing, processing, assembling (i.e., grinding, milling, drying, oil extraction, pressing, construction, carpentry, etc.)		

(continued)

Table 5.2 (continued)

Villages	Mangjing Village (*Blang* minority), Lancang, Yunnan	Daxi Village (*She* minority) Jingning, Zhejiang	Baisha Village Linan, Zhejiang
Proposed Targets on Alternative Energy Development	Potential of unused renewable energy resources, i.e., agricultural by-products, biomass (residues, animal dung and local waste)		
	The synergistic potential of using a 'waste-to-energy service', 'waste-to-wealth' to empower women and marginalised group		
	Fuel switch, improve cooking stove (i.e., biogas, solar cookers)		
	Retro-fit existing built constructs (for energy-efficiency) and manufacturing efficiency (SCP)		
	Between local actors & for a more equitable resource & public service allocation:		
	Develop decentralised energy programs through increased symbiosis integration and synergistic exchange networks		
	Increase local awareness and resources, coherent community consent & attitudes		
	Reinforce continuous adaptive capacity training (particularly awareness, skill sets, some basic business operating concepts)		
	Develop new markets to surrounding counties (energy service, or related cross-sectoral process specialisation)		
	Integrate energy development with multiple renewable resources with micro and small sized entrepreneurship		
	A integrative approach to leverage existing knowledge, assets, resources, culture and habits (such as the combination of the conservation of ethnic culture, natural heritage of tea product and improving their livelihood) via income-generating activities in energy development, green marketing in non-farming business i.e., eco-cultural-tourism, eco-forest tourism		
	Integrate energy services with pooled manufacturing processes, for out-sourced service & distribution—set up centralised community-based production service unit (powered by renewable energy) for key production processes, and employ external training of the key processes and control parameters. In the meantime, reinforce quality & skill set training control to achieve quality, energy and operating efficiency improvements		

Proposed objectives for community transformation	Poverty alleviation and sustainable growth	Poverty alleviation	Consumption pattern change
	Agricultural productivity	To curb growing divide of income or resources allocation.	Growth in Harmony with Nature
	Sustainable forestry.	Agricultural productivity.	Pathway to Eco-Tourism,
	Multiple income (non-farming)	Sustainable forestry	Optimization energy efficiency and resource productivity in local tourist industry
	Jobs creation	Multiple income (non-farming)	To bundle niche tourism-Green Marketing, & renewable deployment (hence Profit Maximization & Branding)
	Entrepreneurship & women empowerment	Jobs creation	
		Entrepreneurship	
		Women empowerment	
		Community training systems	
		Capitalising *She* ethnic legacy for ecotourism—traditional built heritage & abundant natural resources	
		Income growth with bottom-up approach	
Possible issues of energy transition (assumption)	Need for community-based development consent	Full information disclosure to enable community participatory process	Brown baggage
	To reinforce and deepen adaptive capacity training	Inequality of demand (caused by income gap hence consumption demand)	'Over-developed' locked-in facilities and productive assets
		Continuous adaptive capacity training to be strengthened and reinforced	Over consumption from embodied energy, building infrastructures
			Consumption behaviour change

Table 5.3 A decentralised, smaller scale energy alternative and matching business model

Potentials for Decentralized Rural energy access & services (symbiosis combination & local industry synergistic approach)	Output forms & benefits	Opportunities powered by renewable energy and services: @ development programs @ research, demo or replication @ Commercial viability & business opportunities: (in forms of cooperatives , or partner with lead individual, or small or micro business entities)
Solar thermal	Hot water & cooling	The community entity apply for China National program on 'Green energy county development demonstration project'(by 2020) Considering common pool resources, cost appropriation and economy of scale, such as: -Decentralized energy services & waste management centers -Pooled end-use service centers (i.e., community charging, transport, delivery centers) -Distribution / repairing centers for end-use equipments or household appliances (i.e., biogas stove, solar pumps, small bottled bio-fuels lighting) -Pooled quality service, key-process cell production units, outsource service unit, or shelf manufacturing units (to seize energy productivity / quality / productivity / scalability / cost effectiveness / operating efficiency / outsource business service) -Pooled waste treatment / energy service centers -Community-based capacity building & on-job training (i.e., energy-supply related technologies - connecting to the Grid, transformers, distributing, installation mechanical maintenance systems, installing, micro hydro, small diesel engines, mechanics, installing micro wind turbines...etc)
Combined Solar-and PV-panels	Electricity, cross subsidy needed	
Geothermal	Hot water	
Wind energy (Micro turbines)	Electricity	
Hydropower (SHP & MHP)	Electricity	
Gasification and/or combined heat power (CHP)	Gas, electricity, heat	
Heat pumps	Air/air; air/water; water/water	
Burner stove or Direct Combustion	Heat, electricity,	
Organic Rankine heat recovery, CHP, or Fuel Cell system	Electricity , heat	
Heat-recovery from wastewater (heat exchanger)	Heat, hot water	
Biogas, biodiesel, bio ethanol,	Replace fossil fuels	
Improved biomass stoves (replacing traditional stoves)	Productivity, livelihood, social inclusion	
Improved Lighting (replacing candles or kerosene)	Productivity, livelihood, social inclusion	
Improved agro-processing	Productivity, livelihood, social inclusion	
Improved mechanization, motive power & energy services	Productivity & TFP	

present the economic opportunities by tackling barriers to scale in developing economies with a market transition from inclusive business models to inclusive business ecosystems.

Other community-driven schools of thought such as Community-based development (Mathie & Cunningham, 2005), Human development and Result-based Management (Hulme, 2010), and Make Poverty Business (Wilson & Wilson, 2006), work alongside policy initiatives and multinational corporate practice to articulate small-scale, synergistic, enterprise-led development projects at the local level.

According to Engel and Veglio (2010, p. 1):

> Business, as an engine of growth and development, has a critical role to play in accelerating progress towards the MDGs through increasing investment, creating jobs, increasing skills and developing goods, technologies and innovations which can make people's lives better. This includes the private sector's role as a source of capital for developing countries.

Beyond the conceptual framework of cooperate social responsibility (CSR) and 'ethical cooperation', it leaves a further research space to consider; wearing the shoes of the disadvantaged, how to make their job sustainable and meaningful, how to create access to sufficient savings, loans, credits, insurance, educations and other features that provide the capability to enter into profit-making operations like other developed regions. There are targets to be set, questions to be raised and to go through, such as what, how, who, where and how much for the community, economy elites and local leading enterprises alike, who of all can eventually come together to the round table and devise sustainable pathways in order to grow strong together, again with synergistic approaches.

5.9.1 Small Hydropower: Public-Private Partnerships

Given its plentiful wealth of water resources, Jingning picks up the momentum of SH development along with the country's 'China Township electrification programs'—an electricity universal access program for remote areas with the focus on 100 % renewable resources and decentralised systems. Well-known with SH expertise and establishments, Jingning County is celebrated as the 'Hometown of Chinese rural Hydropower' where almost 50 % of the residents enjoy the salient benefits, directly or indirectly, from the local SH investment.[12] According to a local bank survey, the total installed SH in operation, by the end of 2010, numbered 154 establishments with RMB 1.3 billion Yuan investment, whereby another investment (400 million Yuan) to power 70,000 kWh was in the construction stage. The majority of SH investments were structured with public-private partnership (PPP).

PPP describes a public service, primary sectors or private business venture which is funded and operated through the partnership of governments (or institutions) and

[12] For more SHP information, check http://www.hrcshp.org, http://www.inshp.org

one or more private sector companies, through a very detailed contract, inter alia, entails ownership, accountability, legitimacy, effectiveness (manage by results) and interim reports. All actors consent to substantial financial, technical, operational and risk obligation described in the contract. A PPP must be profitable and politically feasible when it comes to transboundary governance or involve the scope ranging from public service, primary sectors to security items such as food, water, energy and land (Schäferhoff, Campe, & Kaan, 2009). In policy and academic debate alike, partnerships are deemed instrumental for international technical transfer, development programs and a fairer social reallocation. It further provides solutions to deadlocked intergovernmental negotiations, ineffective treaties and overly institutional bureaucracies, to power-based State policies, corrupt elites, adulteration of licensing or usage rights (i.e., water or land), and many other real or perceived problems of domestic and global governance alike (Liese & Beisheim, 2011).

Zhao and Zhu (2004) analysed China's SH investment models in PPP. The features mainly include:

- development between enterprises from water resources sector and electric power sector
- development between investment companies and private enterprises at provincial or local level
- development with investment from private enterprises
- foreign invested or joint venture for SH development
- private enterprise financed construction projects of power plants (grids networks are supported through governments or State-owned enterprises).

All above modalities are eligible to be registered as a 'Limited Liability Company' or a 'Share-holding Company Limited' in China. Zhao and Zhu (2004) further concluded that, for SH projects, PPP formation is the more feasible cooperative structure to promote local economies while mobilising capital funding from private sectors.

Following China's 'Big Western Development', 'Jingning SH investment' has reached out to 10 provinces and 51 counties, totalling 225 installations in China's Midwestern areas with the investment amounting to RMB 6.359 billion Yuan. The annual return of investment (ROI) for SH programs is projected at 15–20 % and all commercial loans rendered on SH are in good standings according, to a Lishui City local bank's 2012 interim report.

"Where there are SHPs, there are investors from Lishui; whereas among all the Lishui business people, the 'Jingning army counts the best'." Given this widespread slogan describing that Jingning people are the most successful business people in SH investment, we should not ignore another on the ground reality: namely, that Jingning County is listed as one of China's poorest counties, where the local poor population still receives governmental subsidies for living and development, while an additional welfare package goes to the *She* ethnic group as well. The field visit to Daxi village households revealed that most households use largely fuel wood for daily subsistence, which is suprising considering the widely spread SH program, which was intended for fuel-switch in the first place.

5.10 Discussion and Conclusions

Development is said to be a social equaliser; the study of local energy (SH) development with alternative approaches in the Jingning context suggests further findings of the causal relationship between the existing outcomes and whether community participatory elements[13] were in place and how the elements are carried out, given the fact that poverty still exists in the Jingning area after large-scale SH development already took place. The situation gives a further impetus to locate the true cause of existing poverty and, therefore, the intervention necessary for poverty alleviation. That being said, according to the model provided in this chapter, the next step could be a research project to strengthen community values through cross-sectoral supply chains in energy service, whereby an integrated renewable energy service combining outsourced processing centres can be sketched out for pro-poor and pro-peasant enterprise development.

On the agenda of poverty alleviation from the 'base of wealth' pyramid, Simanis and Hart (2006) state that sustainable poverty alleviation must recognise the poor as central agents in the development partnership process, as key actors in planning and implementing their own development (Binswanger, 2007, p. 2; Dongier et al., 2003). They need to co-create soft infrastructure (service systems) networks to sustain adaptive capacity such as resource efficiency, subsistence farming and agricultural productivity, micro-business innovation, supply chain and service life extension, technology diffusion, local resources utilisation and micro-credits. In poverty alleviation, the major emphasis of development criterion are local knowledge, local context and co-creation processes with the poor that eventually move up the community to develop 'self-help' capabilities alongside exogenous market mechanisms, technical aides and policy supports. Meanwhile, it is also worthwhile to reconsider what policy and local government can do to take the lead, or support in between, or foster throughout the development process of energy poverty reduction.

Over the past decades, multitudes of global crises, humanity issues and climate disasters have shaken up all societies and made us rethink the meaning of 'growth', the pathways and the consequences thereof. Many discussions are directed to balanced growth with 'alternative' approaches which can make possible the necessary transit to sustainable growth. Given that rural electrification mostly reaches areas that meet cost effectiveness of the grid-extension, this leaves the poorest remote communities, often the ethnic minorities, without electricity and modern energy service. In the case of Jingning County, villagers left their beautiful habitats in the mountain and were relocated to the nearest urban cluster for electricity access.

The World Bank (2012) emphasises that issues of community affordability need to be addressed, as once the electricity arrives in a village, the connection charge is the first hurdle that prevents the poor from connecting to the grid. Thus, on account

[13] Community participatory elements are items derived from the CBD (Community-based development) targets such as empowerment, income generation, and shared value.

of commercial viability, most off-grid electrification programs are supported through private sectors, therefore, the social concerns and development models become crucial discussion points for a sustainable rural energy development. This chapter has discussed alternative entrepreneurship business models for poverty alleviation, by improving energy efficiency and resource productivity in a multiple platform with synergistic approach. The human factors, enabling and supportive conditions for the models are not discussed here. Although the chapter is not at all exhaustive, it intends to draw out further discussion on the energy development effectiveness and the associated development modalities.

References

Barnes, D. F., & Floor, W. M. (1996). Rural energy in developing countries: A challenge for economic development 1. *Annual Review of Energy and the Environment, 21*(1), 497–530.

Binswanger, H. P. (2007). Empowering rural people for their own development. *Agricultural Economics, 37*(s1), 13–27. http://onlinelibrary.wiley.com/doi/10.1111/j.1574-0862.2007.00232.x/full

Birol, F. (2007). A place for energy poverty in the agenda? *The Energy Journal, 28*(3), 3. Retrieved http://www.americancorners.or.kr/articlealert/enpq/en7046.pdf

Busan HLF4 Declaration. (2011). Retrieved October, 2012, from http://www.aideffectiveness.org/busanhlf4/en/about/key-documents.html

Cecelski, E., & Unit, A. A. E. (2000). Enabling equitable access to rural electrification: Current thinking and major activities in energy, poverty and gender. *World Development Report, 1*, 2–3.

China Energy Economics Research Institute, Compilation of China policies and regulations on new energy and renewable energy (1986–2011). Economic Management Publishing, ISBN 978-7-2096-1538-6.

China National Bureau of Statistics, Department of Energy Statistics. (2011). *China energy statistical yearbook, Zhongguo neng yuan tong ji nian jian* (Series). Beijing, China: China Statistics Press.

Dongier, P., Van Domelen, J., Ostrom, E., Ryan, A., Wakeman, W., Bebbington, A. et al., (2003). *Community driven development*. World Bank Poverty Reduction Strategy Paper.

Engel, M., & Veglio, E. (2010). A business view on development. *European Business Review Online*. http://www.europeanbusinessreview.com/?p=2339. Retrieved March 4, 2014.

Gradl, C., & Jenkins, B. (2011), *Tackling barrier to scale: From inclusive business models to inclusive business ecosystems*. CSR Initiative at the Harvard Kennedy School. http://www.hks.harvard.edu/m-rcbg/CSRI/publications/report_47_inclusive_business.pdf. Accessed Dec 2012.

Harvard CSR Initiative. (2007). Retrieved July, 2012, from http://www.hks.harvard.edu/mrcbg/CSRI/publications/report_29_Harvard %20EO%20Dialogue%20Summary%2020071018.pdf

Hulme, D. (2010). Lessons from the making of the MDGS: Human development meets results-based management in an unfair world. *IDS Bulletin, 41*(1), 15–25.

International Energy Agency. (2011). *World energy outlook* (Released – November 9, 2011). ISBN: 978-92-64-12413-4. http://www.worldenergyoutlook.org/publications/weo-2011/

International Energy Agency. (2012). *Medium-term coal market report*. Retrieved January 2012, from, http://www.iea.org/newsroomandevents/news/2012/december/name,34467,en.html

International Energy Outlook. (2011). *U.S energy information administration. World total primary energy consumption by region*. Reference case, 2006–2035. Retrieved January, 2012, from, http://www.eia.gov/oiaf/aeo/tablebrowser/0-subject=IEO2011&table=1IEO2011&region=0-0&cases=Reference-0504a_1630

Jiang, L., & O'Neill, B. C. (2004). The energy transition in rural China. *International Journal of Global Energy Issues, 21*, 2–26.

Kandachar, P., & Halme, M. (2008). Introduction. Farewell to pyramids: How can business and technology help to eradicate poverty? In P. Kandachar & M. Halme (Eds.), *Sustainability challenges and solutions at the base of the pyramid: Business, technology and the poor* (pp. 1–27). Aizlewood's Mill, UK: Green Leaf Publishing. ISBN 978-1-906093-11-2.

Kaygusuz, K. (2012). Energy for sustainable development: A case of developing countries. *Renewable and Sustainable Energy Reviews, 16*(2), 1116–1126. http://www.sciencedirect.com/science/article/pii/S1364032111005491

Liese, A., & Beisheim, M. (2011). Transnational public-private partnerships and the provision of collective goods in developing countries. In T. Risse (Ed.), *Governance without a state? Policies and politics in areas of limited statehood* (pp. 115–143). New York: Columbia University Press.

Lu, H., & Li, L. (2006). An empirical analysis of the impact of the impact of farmers' income level on the household energy consumption structure in the countryside. *Finance and Trade Research, 3*, 28–34.

Luo, G., & Zhang, Y. (2008). Analysis of energy consumption in rural areas of *China. Agricultural Science Bulletin, 24*(12).

Mathie, A., & Cunningham, G. (2005). Who is driving development? Reflections on the transformative potential of asset-based community development. *Canadian Journal of Development Studies/Revue canadienne d'études du développement, 26*(1), 175–186.

McKinsey Quarterly. (2011). Another oil shock? Retrieved December, 2012, from https://www.mckinseyquarterly.com/Another_oil_shock_2873

NDRC. (2011). *China's twelfth five year plan* (2011–2015). National Development and Reform Commission (European Delegation to China, Trans.). Retrieved February 28, 2014, from http://www.britishchamber.cn/content/chinas-twelfth-five-year-plan-2011-2015-full-english-version

Ness, D., & Vogiatis, N. (2013). A clean 'energy plus' rural economy in China. In M. Liu, M. Henry, & H. Huang (Eds.), *Renewable energy in China: Towards a green economy*. Hong Kong/Singapore: Enrich Professional Publishing.

Pachauri, S., & Spreng, D. (2004). Energy use and energy access in relation to poverty. *Economic and Political Weekly*, 271–278.

Reddy, A. K., Annecke, W., Blok, K., Bloom, D., & Boardman, B. (2000). *Energy and social issues, 39–60. World Energy Assessment: Energy and the challenge of sustainability*. New York: United Nations Development Programme. http://www.undp.org/content/dam/undp/library/Environment%20and%20Energy/Sustainable%20Energy/wea%202000/chapter2.pdf. Accessed Nov 2012.

Schäferhoff, M., Campe, S., & Kaan, C. (2009). Transnational public-private partnerships in international relations: Making sense of concepts, research frameworks, and results. *International Studies Review, 11*, 451–474. doi:10.1111/j.1468-2486.2009.00869.x.

Shen, L., Liu, L., Yao, Z., Liu, G., & Lucas, M. (2010). Development potentials and policy options of biomass in China. *Environmental Management, 46*(4), 539–554. http://link.springer.com/article/10.1007/s00267-010-9476-4#

Simanis, E., & Hart, S. (2006). Expanding possibilities at the base of the pyramid. *Innovations: Technology, Governance, Globalization, 1*(1), 43–51.

Simon, D. (1997). Development reconsidered: New directions in development thinking. *Geografiska Annaler: Series B, Human Geography, 79*(4), 183–201.

UNESCAP. (2010). *Eco-efficiency indicators: Measuring resource-use efficiency and the impact of economic activities on the environment*. Bangkok, Thailand: Environment & Development Division. http://www.unescap.org/publications/detail.asp?id=1386

von Bertalanffy, L. (1972). The history and status of general systems theory. *The Academy of Management Journal, 15*(4), 407–426.

von Weizsacker, E. U., Hargroves, C., Smith, M. H., Desha, C., & Stasinopoulos, P. (2009). *Factor Five: Transforming the global economy through 80% improvements in resource productivity*. London: The Natural Edge Project/Earth Scan.

Wang, X., & Hu, X. (2010). Factors influencing rural household energy consumption [J]. *Transactions of the CSAE, 26*(3), 294–297.

Wang, J. J., Jing, Y. Y., Zhang, C. F., & Zhao, J. H. (2009). Review on multi-criteria decision analysis aid in sustainable energy decision-making. *Renewable and Sustainable Energy Reviews, 13*(9), 2263–2278.

Warr, B., & Ayres, R. (2006). REXS: A forecasting model for assessing the impact of natural resource consumption and technological change on economic growth. *Structural Change and Economic Dynamics, 17*(3), 329–378. http://www.sciencedirect.com/science/article/pii/S0954349X05000329

Wilson, C., & Wilson, P. (2006). *Make poverty business*. Sheffield, UK: Greenleaf.

World Bank. (2012). *The social impact of rural electrification*. Retrieved October, 2012, from http://siteresources.worldbank.org/EXTRURELECT/Resources/exec_summary.pdf

Yumkella, K., Bazilian, M., & Gielen, D. (2010, April). Energy for all. *Making it: Industry for development* (pp. 22–29, Vol. 2). http://89.206.150.89/documents/congresspapers/325.pdf. Accessed Nov 2011.

Zhao, J., & Zhu, X. (2004). *Private participation in small hydropower development in China – Comparison with international communities*. UNHYDRO 2004. http://www.un.org/esa/sustdev/sdissues/energy/op/hydro_zhao_english.pdf

Chapter 6
Transforming Rural Communities by Synergistic, Integrated and Inclusive Planning of Services and Infrastructure

6.1 Introduction

In Chap. 2, social entrepreneurship was described as a process of creating social value by offering products and services in new ways, involving the innovative use and combination of resources. It highlighted the need for infrastructure and other services to support social entrepreneurs and enterprises, such as energy, water, and transport to deliver goods to markets. In this regard, the remoteness of many villages poses special challenges and has led to many villagers being resettled in urban areas. Protecting existing culture, architecture and environments was identified as another major challenge for some villages. This chapter further develops these themes, focussing on innovative ways to provide services in an integrated, symbiotic manner, whilst sharing and optimising the use of local resources and capabilities.

Previous research has highlighted the value of such symbiotic approaches, where the diverse views and values of multiple stakeholders are brought together to achieve community transformation, supported by integrated and connected infrastructures (Devisscher & Mont, 2008; Ness & Xing, 2010). In addition, 'solution-oriented partnerships' (SOPs) have been proposed to capture synergism among elements, thereby driving the co-production of service and physical infrastructure solutions and achieving 'creative holism' (Jackson, 2006; Morelli, 2006).

Based upon fieldwork in remote mountain regions of south-west and south-east China, the authors seek to test the proposition that, using the method, the limited local resources, assets, infrastructure and capabilities of various actors within rural communities may all be marshalled to serve a number of community enterprises and thereby lead to stronger communities. In addition, villages may be provided with infrastructure that is not only efficient and effective, affordable and equitable, but also appropriate and sensitive to cultural and natural settings. It is especially important that opportunities are found for communities and their

Y. Zhu et al., *Transforming Rural Communities in China and Beyond*,
DOI 10.1007/978-3-319-11319-7_6

entrepreneurs to plan, construct, maintain and manage their own local infrastructure, supported by training and capacity building, thereby creating local infrastructure enterprises, boosting jobs, incomes and confidence. As the World Bank (2012) stated, "...jobs that do the most for development can spur a virtuous cycle".

The purpose of the research is explained in Sect. 6.2. The theory and literature related to symbiosis, systems theory and product-service systems (PSS) is then outlined in Sect. 6.3, with the notion that creative synergies may enable more services and infrastructure to be provided with less resources, cost, and pollution. The extent to which this has been applied to community transformation—and especially rural villages—is examined. A strategic planning method for synergistic community transformation, which uses PSS principles in conjunction with soft systems methodology (SSM), is outlined in Sect. 6.4. The method is then applied to the case study communities in rural China, as described in Sect. 6.5. Finally, Sect. 6.6 briefly comments on the findings and their implications for synergistic or symbiotic sustainability, identifying areas for further research.

6.2 The Research

6.2.1 Research Background

The proposition examined by this chapter is that an interconnected and integrated infrastructure system may better support community enterprise development, especially when it is inclusive, appropriate to local culture and natural settings, affordable and creates enterprise and employment opportunities.

6.2.2 Research Questions

The following are the key research questions:

(a) What is the theoretical and empirical basis that connected-up and synergistic approaches deliver improved outcomes?
(b) Can integrated, connected-up approaches enable infrastructure to support community social entrepreneurs and enterprises more efficiently and effectively?
(c) Will such approaches enable infrastructure to be more affordable, equitable, inclusive, and appropriate to cultural and natural settings?

These questions will be examined later in this chapter by application of a strategic planning method to case study villages.

6.3 Theory

6.3.1 Synergistic Community Planning

As described by Corning (1998, p. 3), "synergy is...a ubiquitous and fundamentally important aspect of the natural world", related to the effects produced by things that operate together (parts, elements or individuals). In essence, "the effects produced by wholes are different from what the parts can produce alone". Similarly, 'symbiosis' is a biological term defined as "a close association of two different animal or plant species or groups living together to their mutual benefit" (Sinclair, 1995); the parts of natural ecosystems are intrinsically connected and symbiotic. Industrial ecology seeks to apply the principles of natural ecosystems to industry, engaging traditionally separate industries in a collective approach, whereby the 'waste' from one enterprise provides the raw materials for another, forming closed loop, interconnected and symbiotic systems (Tilley, 2003; Chertow, 2000).

Pro-poor eco-settlement action research undertaken within the Philippines identifies "holistic and synergistic, community-based solutions for issues such as site development, infrastructure and housing, health and food security, income generation, environmental risk reduction, as well as community empowerment and resilience" (UNESCAP, 2011, pp. 165–175). It proposed 'synergistic development scenarios' that, through emphasising integration and interlinking, can uncover complementarities leading to 'win-win-win' solutions encompassing the economic, social and environmental dimensions of sustainability.

The symbiotic, integrated ecological development model of Gaviotas, Columbia, provides a further inspiration. The community is pursuing integrated strategies and solutions that mimic the sustainable cycles and interconnections of natural ecosystems whereby, via a systems approach, all problems are being tackled concurrently in a linked manner (Nicholls, 2006). Similarly, Takeuchi, Namiki, and Tanaka (1998) designed models of ideal eco-villages for revitalising rural Japan, with interconnections between enterprises, housing and other facilities and the natural environment, with low inputs and sustainable material flows. For example, animal and human waste is used for producing methane gas, which is converted into energy; agricultural waste is fed into a compost plant that feeds cropland, which in turn provides food for local residents; grey water is passed through wetlands before being discharged to the river; a 'disposer' is used for recycling resources and improving living standards.

Arguably, the creation of stronger communities, with economic participation and access to services at the core, will be more effective and efficient when a connected-up and holistic view is taken of services and supporting infrastructure, with all actors working in unison towards common ends. This connected-up, integrated thinking may also enable services, products, assets and infrastructures to be coordinated across sectors and to be mutually reinforcing (see Australian Asset Management Collaborative Group, 2011).

6.3.2 Systems Thinking

Systems thinking helps uncover the interconnections between ideas thought relevant to a situation, rather than considering these in isolation. For example, echoing the work of Ness (2008), Joham, Metcalfe, and Sastrowardoyo (2009, p. 788) explained:

> ...thinking about a road project in a developing country in terms of its connectivity (relationship) to other projects may result in joint infrastructure between the road and water or solar energy catchments'. Building a new road will help get crops to market. The water catchments project might be to grow crops and the solar energy might be needed to process the crop somehow. These ideas can be connected.

Alexander (1964) pioneered the idea of connecting ideas via a network, examining the connected components of an Indian village including agriculture, animal husbandry, employment, water, transportation, soils, education, health, material welfare, social forces, religion and the like. 'Functioning linkages', 'synergies and co-benefits' may be uncovered, leading to 'systems innovation'.

As noted by Dewey (1910) and Joham et al. (2009), taking a wider perspective by 'zooming out' from a particular system element and its boundary enables interrelationships with other elements to be more easily identified. For example, we may find connections between various elements that contribute to a rural community's economic development: the agricultural produce, means of transporting this to markets, marketing the produce, training the operatives, access to water and energy, waste and wastewater, and the like. On the other hand, if we wish to analyse specific elements of the system more closely, we may 'zoom in' for detailed investigation, such as the type of agricultural products and their possible diversification or more viable alternatives.

6.3.3 Product-Service Systems (PSS)

Vargo, Maglio, and Akaka (2008) explained, a 'service system' is a configuration of resources (including people, information and technology) connected to other systems by value propositions, while 'service science' is the study of service systems and of the co-creation of value within a complex configuration of resources.

A PSS is a particular kind of service system, including products and services delivered to customers, the networks of actors, supporting infrastructure and other enabling mechanisms. Mont (2004, p. 259) viewed a PSS as "a system of products, services, networks of actors and supporting infrastructure that continuously strives to be competitive, satisfy customer needs and have a lower environmental impact than traditional business models". Reflecting the systems theory described above, Devisscher and Mont (2008, p. 265), emphasised that "it is important to consider the function that is delivered to the customer from a systems perspective and *optimize the entire system and not a single part of it*", from economic, social and environmental perspectives (italics added).

Co-development and co-delivery of PSS solutions has notable power for service system innovation; harnessing the complementary competencies, resources and capability of different organisations makes it possible to produce services beyond the scope and capability of individual organisations (Partidário, Lambert, & Evans, 2007; Pawar, Beltagui, & Riedel, 2009). By this means, additional value may be co-created (Vargo et al., 2008). Jégou and Joore (2004), Manzini, Collina, and Evans (2004) and Krucken and Meroni (2006) discussed the concept of a 'solution-oriented partnership' (SOP) and methods to develop PSS solutions based on SOP. This involved a systems approach for formation and collaboration of networks of partners with a common vision and the capacity to develop context-focussed, sustainable solutions involving 'system resource optimisation' (Morelli, 2006). In considering potential benefits of PSS applications to developing countries, Manzini and Vezzoli (2002) stated: "a PS systems approach could reach and provide a widespread higher level of well-being or utility at lower cost, because of the higher system efficiencies". Similarly, the 'One Village, One Product' (OVOP) scheme began as a way of developing products or services within rural communities by adding value to locally available resources and diversifying agricultural practices and products. The idea was for the world's poorest communities to take control of their own lives by creating competitive products made from local materials and knowledge, and provide equipment, human resource development and marketing expertise to facilitate the process, encouraging 'entrepreneurial independence' (Wahlin & Natsuda, 2008).

Consistent with and extending such thinking, Ness and Xing (2010) proposed a typology and analytical framework for implementing PSS solutions to transform communities of low social-economic status (Fig. 6.1). Focussed on the community and its transformation, the framework includes PSS elements such as *products* (consumer, commercial and capital goods); *services* such (as technical, business and financial supports); *enabling mechanisms* (such as infrastructure, resources, policies, technologies, financial instruments, markets and trade); and *actors* (such as producers and their cooperatives, the local community/local government, non-government organisations (NGOs), financial and other institutions).

The typology and framework sought to extend the work of Devisscher and Mont (2008) who analysed the case of coffee production in the Yungas region of Bolivia from a PSS perspective. They demonstrated that a shift to shared ownership of equipment via a cooperative (CENCOOP) had economic, environmental and social benefits, improving productivity, social livelihoods and confidence within the community. The authors found that the main driver for the formation of CENCOOP and its shared ownership was the economic interest of members to be more competitive in the international coffee market. The solution involved a shift from individual product ownership to a managed arrangement of utility provision via a mix of products and services. The elements of the PSS included products, services, infrastructure, and a network of actors. In terms of products, two main products were considered: firstly, the coffee, and secondly, the equipment used for its processing. Services included support services at the processing point, production and capacity building services, and promotion and marketing services. Infrastructure included

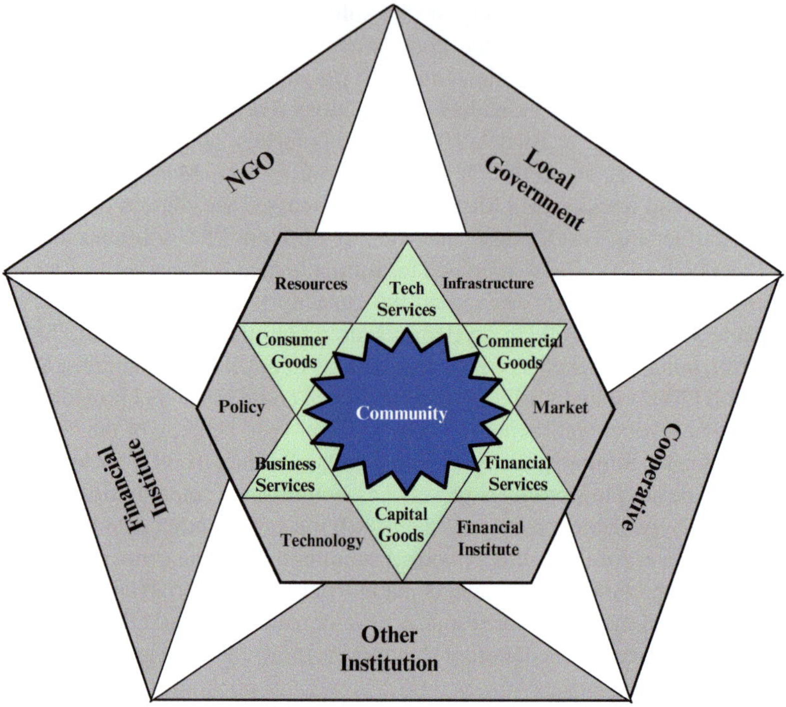

Fig. 6.1 A framework for community transformation based on PSS (Source: Ness & Xing, 2010)

the processing facility itself, transportation facilities, solid waste treatment infrastructure, waste water treatment, promotional facilities such as a coffee bar and museum, and intangible infrastructure in the form of international market channels. In addition to the coffee producers, retailers and consumers, the shift to a collective sharing system brought other actors into play including certification companies, NGOs for financial, institutional and technical support, brokers and coffee businesses.

Thus, the combination of products and services within a system has the potential to provide multiple services from limited products and resources by means of symbiotic approaches and system resource optimisation.

6.3.4 Doing More with Less

Von Weizsacker et al. (2009, p. 34) advocated a 'whole system approach to resource productivity', which is the economic output or value added per unit of resource use (Ritthoff, Holger, & Liedtke, 2002). By considering the interconnections between systems, seeking solutions that address multiple problems at the same time, cumulative and synergistic improvements in resource productivity can be delivered; that is, more services or outputs, with less material input (Schmidt-Bleek, 2000). UNESCAP (2011) has illustrated the application of these principles to urban infrastructure planning via 'orchestration of sectors' and 'harnessing existing assets', creating more value for citizens whilst reducing the use of resources and the production of waste and pollution. Consistent with the concept of eco-efficiency developed by the private sector in the early 1990s (WBCSD, 2000), "an eco-efficient approach to urban infrastructure development seeks to highlight ...multiple returns and help prioritise policy options that lead to win-win scenarios" (UNESCAP, 2011, p. 27). The principles of eco-efficient infrastructure planning can also be adapted for rural contexts, focussing not only on improvements that yield parallel economic and environmental benefits but also those that are socially advantageous. Promoting shared and multiple uses of assets through co-location of services are other ways to use assets and resources more efficiently and sustainably (Office for Infrastructure Development, 2005). Thus, the literature and case studies indicate that connected-up, symbiotic approaches may be more effective and efficient than those that are disconnected and isolated.

6.4 A Strategic Planning Method

The UNESCAP (2011) has outlined a strategic approach to infrastructure planning and development, which involves mobilising different views and resources towards a common vision, goals and objectives, with careful consideration of various 'win-win' solutions. UNESCAP's four-stage, ten-step approach addresses four essential questions: Where are we now? Where do we want to go? How do we get there? Are we getting there? (UNESCAP, pp. 59–61). Coupled with processes from the literature, this approach and its associated steps have been adapted for strategic planning of services and infrastructure for the communities studied in China, as follows:

The key stages associated with the method are:

- *Understanding the problem situation and context (Where are we now?).* This includes identification of stakeholders, understanding their views, and understanding environmental, economic, cultural, and social-political characteristics of the situation; a profile of the community may thus be developed. The perspectives and boundaries of the study may also be adjusted i.e., zooming out for synthesis or zooming in for closer analysis (Dewey, 1910; Metcalfe, 2007).
- *Transformation goal statement (Where do we want to get to?).* This states the desired purposes and outcomes of transformation agreed upon by the actors within the community. According to Joham et al. (2009) who applied the methodology to a village, the 'transformation statement' seeks to change some defined input into some defined output, whilst accommodating diverse viewpoints and world- views. For example, an economic world-view might be encapsulated in a transformation statement such as 'From a village which has a low GNP to one with the appearance of a higher GNP within 5 years' (Joham et al., 2009, p. 791).

- *Solution oriented partnerships (How do we get there?).* This involve groups of heterogeneous stakeholders (such as actors in the community, corporative entities, government agencies, financial institutions, NGOs, etc.) that have similar interests for value co-production (Ramirez, 1999). Such partnerships represent the competencies of the participating entities in terms of the goods and hardware that they can provide, the kinds of services that they can offer, and the resources and infrastructures that are available for solution development.
- *Product-service modes (What do we need to get there?).* These represent a cluster of possible actions or solutions to address the defined 'problem situation' and to achieve the desired transformation. A PSS mode consists of a scenario (indicating a theme or strategy for change), a task force (i.e., a group of participants that collaborate under the particular theme), and the product-service (P-S) configurations that represent the specific technical and service solutions related to the theme.

Based on a holistic view, different PSS modes can be interconnected through synergies among their scenarios, task forces and P-S configurations, with overlaps involving the actors, their capabilities, resources and infrastructure involved, as illustrated in Fig. 6.2.

The creative process of uncovering connections and synergies between various scenarios and PSS modes is aided by techniques within system theory, such as zooming out our system lens and taking a wider 'view from the top' perspective.

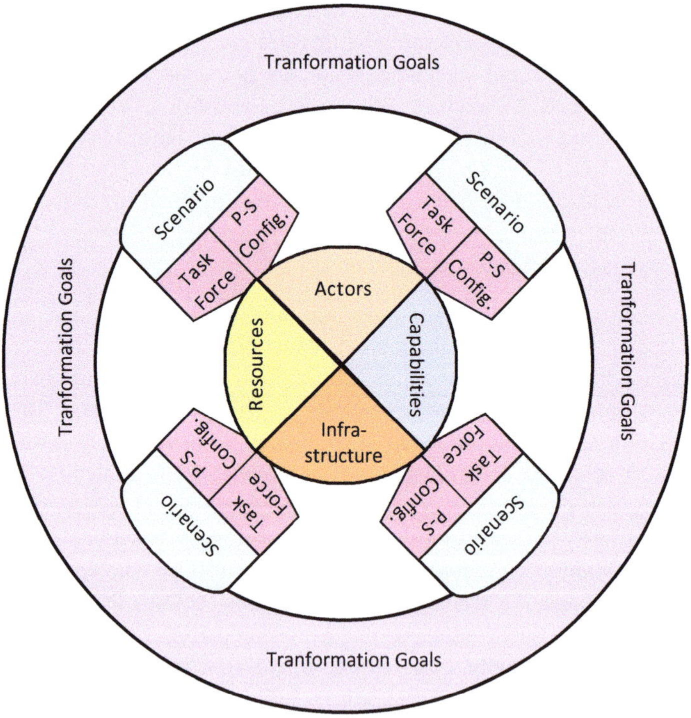

Fig. 6.2 A holistic view of connections between PSS modes (Xing, Ness, & Lin, 2013)

Such interconnections are often important for the efficiency of resources and infrastructures and essential for the viability of the solutions, especially in remote mountain areas where limited resources need to be marshaled and shared for the common good—as in the case of the OVOP scheme (Wahlin & Natsuda, 2008).

6.5 Applying the Method to Rural Villages in China

6.5.1 The Case Studies

This section explains the application of the method by the research team to the case study villages in rural China, which were largely described in Chap. 1. For the purposes of this chapter, the case study villages include:

- *Mangjing village*: A remote *Bulang* minority community within the mountains of Lancang County, Yunnan, south-west China.
- *Daxi village*: Located in the mountains of Jingning autonomous County, which is a *She* minority community—designated by the Chinese Government as poverty stricken—located in the mountains of Zhejiang, south-east China.
- *Jingning village*: A *She* village (being relocated) near Jingning City, Jingning autonomous County.
- *Baisha village*: Located in the mountains of Lin'an County within Zhejiang Province and achieved recognition as a model eco-civilisation village in 2011.

6.5.2 The Findings

6.5.2.1 Understanding the Problem Situation and Context (Where Are We Now?)

As Checkland (2001, p. 71) said, "soft systems methodology articulates a process of organised finding out about a problem situation". In line with that approach, the nature and extent of the context, its location and its challenges has to be understood and investigated with regard to the social, economic, environmental, cultural and even political characteristics.

Some key aspects of the case study communities are now highlighted.

Mangjing Village (*Bulang* Ethnic Minority)

This series of villages on Jingmai mountain, in the far south-west of Yunnan, is home to the ancient *Bulang* ethnic minority, who have grown tea in the area for several thousand years and display colourful costumes and rich traditions. Tea plants are integrated with forest canopies as 'tea forests' and demonstrate co-existence and synergies between production of tea and conservation of forests and biodiversity, with over 240 species. As Liang (2010) explained:

> The forest canopies suppress weeds and provide fallen leaves and litter as 'green manure'. The canopy trees also protect soil from erosion on the steep slopes.....forest canopies nurture natural enemies such as spiders, birds and wasps that control pests in the tea gardens.

The Mangjing tea forests therefore exemplify natural symbiosis, while the local people harness the eco-services of rich biodiversity to support their tea production, without the need for chemical pest control and fertilisers.

A brand 'Abaila' (Tea Soul) was registered, and a mantra of 'one cooperative one brand' was created. The Ford Foundation also launched a program to gather local *Bulang* villages together for tea production in the name of both poverty alleviation and cultural revitalisation (Hung, 2012). However, this did not survive the test of time and a number of disparate economic co-operatives and brands have since been formed—such as by village capable person B. It is a challenge for the village chief (social entrepreneur A) to draw these initiatives together for the common good.

While a private tea company ('Brilliant') located on Jingmai mountain employs many villagers, improving their incomes and security, the villagers lack entrepreneurship opportunities and control of their own land and destiny.

Inadequate infrastructure is a factor inhibiting village tea drying and livelihoods. Several hundred fragmented households use one electricity transformer, there is insufficient water for tea production and drinking, and there are natural obstacles and great distances associated with transporting products to markets and attracting tourists. Houses are generally constructed of timber, due to Government controls over the tourist site, but villagers prefer to construct in cheaper and more durable concrete.

Daxi Village (*She* Ethnic Minority)

The community is located in a beautiful mountain setting within Jingning Autonomous County, with forests, streams, waterfalls, walking tracks, the historic Huiming temple and a geological museum. The ancient *She* culture, exemplified by the iconic *She* multipurpose temple bridge in Daxi village, is celebrated in a museum in Jingning City, and the area is known for its many natural or 'wild' foods. Micro-hydro technology provides up to 10 % of energy in Jingning. However, the village, region and Jingning County in general are at the developmental stage, facing the challenge to increase tourist numbers and expand enterprises while preserving the natural and heritage assets. With considerable transport investment in Jingning and Zhejiang (such as tunnels through mountains), access to the markets of Shanghai, Hangzhou and other areas has much improved and exceeded access in the region inspected within Yunnan. The village chief (C) emphasised that the community was at a critical juncture, with opportunities now needing to be seized to increase prosperity and overcome poverty.

Jingning Village

The researchers inspected a village near Jingning City that, whilst accessible by road and at the base of the mountain, was being closed and the inhabitants resettled in nearby urban areas. However, there is a risk that ancient *She* ethnic culture and heritage, which is so much allied to mountain settings, may be diluted or even lost. While the traditional housing lacked modern conveniences and amenities, the village exhibited many traits of an eco-village, with bamboo watering systems for crops, and even a stream, with fish, flowing through the rear of a household kitchen.

Baisha Village

This village is known for transforming its economy from cutting trees to one based upon forest conservation and accredited organic forest products - as the professional adviser (social entrepreneur F) expressed: 'From people who cut trees to people watering trees'.

As shown during an inspection of the associated Dong Keng village and environs, the forest is an example of resource productivity via a symbiotic, synergistic system, providing multiple benefits via its three layers, in a similar manner to the Manjing tea forests. The upper layer forestry canopy provides shade, suppresses weeds, and leaf litter; the middle layer offers the potential of non-wood agricultural specialisation with cash crops such as tea, pecan and walnut; while the bottom layer's fertile soil can produce bamboo shoots and vegetables throughout different seasons.

Over 65 % of Baisha's GDP is from tourism, although interviews revealed that this is seasonal. The professional adviser (F) pointed out that Baisha, while recognised as an eco-village, faces over-development that threatens the fragile environment upon which ecotourism and natural products depend and increases inequity. Some inhabitants have much improved incomes and tend to squander this on large houses that are not in keeping with the landscape, while tourist hotels grow in an untrammelled manner and tourist waste looms as an increasing problem. Similar to Daxi village, it was recognised that villagers require training to improve their capacity to manage their affairs, accompanied by expert advice, such as in sustainable design.

As noted earlier in Chap. 4, Lin'an already has an international reputation in bamboo plantations, with recognised expertise in bamboo carbon sequestration.[1]

---

[1] See www.hzbamboo.net

6.5.2.2 Transformation Goal Statement (Where Do We Want to Get to?)

Understanding and learning about the problem lead to deliberate actions to bring about improvements, as explained by Checkland (2001). SSM can be used to bring forward the various world-views of stakeholders, with the ideas network approach outlined by Metcalfe (2007) being an especially useful method of finding connections and, desirably, accommodation between various viewpoints of desired changes. The 'transformation goal statement' should encapsulate some systems of purposeful activities relevant to the problem.

Mangjing Village

As was noted in Chap. 2, the village chief (rural social entrepreneur A) is seeking to bring together the various tea enterprises and cooperatives within Mangjing village under a common brand and shared processing facilities. In order to lead his hesitant people progressively towards this goal, he is investing his own funds into the facility, with his own mortgage. He expressed the vision that Mangjing should grow and compete with the neighbouring Brilliant tea company, enabling the villagers to prosper and determine and control their own destiny whilst preserving their ancient culture and tea growing traditions. Whilst this vision does not yet represent the views of the entire community, it is assumed to be the transformation goal.

Daxi Village

The chief of Daxi village (social entrepreneur C) has a clear vision, as described in Chap. 2, involving balanced development: "The priority is not just increasing income, but also investment in hard and soft infrastructure (including education), so that people are better prepared and have improved capabilities". His vision also encompasses preserving and developing traditional culture and architecture, ecological development in harmony with the ecosystems and landscape, promoting the natural foods and medicines, increasing household income, and developing a beautiful and liveable village. He highlighted the need for 'integrated strategies' and a marketing strategy to attract more visitors. We can imagine a vision and brand for a 'unique Jingning' as a 'cultured, clean, competitive, confident country', integrating the various elements described by the village chief.

Jingning Village

The village near Jingning City has been largely abandoned, although several older residents remain. Whilst it seemed to lack a village governance structure, the researchers recognised its potential to be transformed into an ecological village and tourist development if it can incorporate modern amenities and infrastructure. Its character of being harmonious with nature, and its cultural assets and architecture can provide a sound base upon which to build.

Baisha Village

Led by the village chief (social entrepreneur D), the village has already undergone a transformation from an economy based on cutting the forest to one based on its conservation and cultivation, with the assistance and advice of a professional person (social entrepreneur F) with expertise in organic production and forestry. A further transformation goal for Baisha now emerges from the current challenges. It is clear that household construction and tourist development (social entrepreneur E) needs to be more in harmony with the natural setting, that is, eco-development. In addition, systems need to be put in place to reduce and recycle waste, a point raised by the adviser (F).

6.5.2.3 Solution Oriented Partnerships (How Do We Get There?)

Along with the problem identification and transformation goal definition, a solution oriented partnership may be formed with (desirably) a common understanding of *what* is needed to engage with the identified problems and *how* to achieve the desired transformation, utilising the community's inherent strengths and building upon their joint competencies and resources.

In the case of the villages mentioned above, the village council or cooperative is an example of an overriding SOP (such as for tea products and services), whilst various other specialised cooperatives may also exist, such as in the case of Daxi village. Various members of particular villages may have specific capabilities and capacity in various forms of infrastructure (e.g., energy, water and waste) and construction, with some displaying leadership and social entrepreneurship skills.

The SOP may also involve partnerships involving community actors, local governments and the private sector, as in Public-Private Partnerships (PPPs), as described in Chap. 5 in relation to energy services.

Mangjing Village *(Bulang)*

In addition to the village chief (A), who was seeking to draw together all groups and households, the village capable person (B) led a tea cooperative and retained *Bulong* culture via a women's song and dance group, exhibiting capabilities and capacity. In addition, she displayed knowledge in construction, was generating lighting from animal waste derived methane, and recycled water from washing tea leaves to create steam for drying the leaves. She also used tea leaf residues for producing handicrafts.

The private Brilliant tea company was using coal energy for drying tea leaves, but was directing this through water so that emissions were cleaner. It was also pursuing solar technology, and this knowledge could be invaluable for the neighbouring *Bulang* tea producers.

Daxi and Jingning Villages

Interviews with villagers and officials revealed that the *She* community has considerable capabilities and capacity in natural, organic foods and products, derived from its traditional close and spiritual association with mountains and forests. Despite being threatened by the closure of mountain villages, effort was also being made to preserve traditional architecture and costumes, including fine cultural and geological museums. However, community leaders such as the Daxi village chief (C) recognised increased management, business and marketing skills are required to capitalise on the varied and rich natural and heritage assets, with the recent remarkable improvements in transport infrastructure affording access to markets for tourism and natural products.

As noted in Chap. 2, Daxi villagers have formed different cooperatives with specialisations in particular commodities, such as red rice, tea and bamboo. In addition, the community jointly developed capabilities in micro-hydro technology, which is exported to other parts of Zhejiang and China via local infrastructure enterprises, further boosting local incomes. Villagers were also being encouraged to develop holiday houses and hotels, although capabilities appeared to be at an embryonic stage.

Baisha Village

As revealed from interviews and meetings, the village council and community included a number of members with expertise in holiday houses and tourism. One successful entrepreneur (E) who ran multiple businesses was the acknowledged leader of holiday house ecotourism. Another villager with building expertise undertook much of the construction of tourism and other infrastructure. Such capable individuals could well form the nucleus of a SOP focussed on more appropriate and harmonious design and construction, drawing upon the know-how of the nearby Dong Keng village and the assistance of external advisers such as architects with expertise in sustainable and symbiotic design. This could mirror the village's previous transformation, where the professional adviser (F) played such an important facilitation role.

6.5.2.4 Product-Service Modes (What Do We Need to Get There?)

Such partnerships, coupled with the product and service competencies of the actors as well as physical and non-physical resources involved, may be manifested in P-S modes', consisting of scenarios or transformation themes, task forces and P-S configurations, as illustrated earlier in Fig. 6.2. A number of scenarios and modes can be conceived, each reflecting the world-views of the various actors and partners, employing their competencies and resources, and consistent with the transformation goals. Although further research is required to explore in more depth the views and ideas of the villagers themselves, the following are some potential scenarios largely developed by the research team, based upon the fieldwork and literature.

Mangjing Village

One 'scenario' or transformation theme is based upon common branding of the Mangjing tea and a combined, shared processing facility, as sought by the village chief (A). This will hinge upon villagers recognising the mutual benefits of an overall brand for their various initiatives, and becoming active and financial partners in the cooperative or task force.

Manjing village could learn from the CENCOOP coffee cooperative, Yungas, Bolivia, which was analysed by Devisscher and Mont (2008). To improve the quality of their coffee and its market competiveness, five coffee cooperatives decided to join operations in a task force, shifting their individual processing system to a collective one, whereby four micro-scale plants are owned and shared by all CENCOOP members. This resulted in better and more professional equipment, making the P-S configuration more efficient, less resource and time intensive and leading to a more price and quality competitive product. In addition, CENCOOP offered a number of other services to its members including advice on coffee growing practices, water conservation, access to wider markets, organic certification and eco-labelling.

As unreliable energy supply was highlighted as a challenge in Mangjing, a similar task force could also develop and share products and technologies such as solar and biomass energy to support the drying and processing of tea (the 'service'). Building upon the know-how and initiatives of individual entrepreneurs, the production of methane and energy from tea plantations, animal and human waste and biomass could be a viable option, as was discussed earlier in Chap. 5. Similarly, water from washing tea leaves can be recycled to create steam for drying the leaves. These are all examples of P-S configurations.

Another scenario and mode may involve culturally appropriate products such as architecture and crafts to support ecotourism services. Whilst it is recognised that construction using local materials (e.g., timber, bamboo) is more expensive than concrete, the additional cost may be offset by opportunities for reviving unique local trades and crafts and the employment benefits this may bring. External tourism providers could be co-opted to the task force to assist with finance and investment, also training in business, hospitality and marketing of the ecotourism products and services. Similarly, as one interviewee suggested, a *Bulang* culture museum could form part of the tourism package and partnership (similar to the *She* museum, Jingning County).

Trading of forest property rights and carbon credits may constitute another major opportunity for the Mangjing community to boost incomes, employing the methods described in Chap. 4. Partnership agreements with afforestation companies may also include the provision of much needed infrastructure such as roads and other facilities.

Daxi and Jingning *She* Villages

Similarly, increased ecotourism is one scenario for a more prosperous Jingning, utilising its pristine natural features, clean air, foods and forests, with facilities in harmony with the ecosystems and landscape, accompanied by 'beautiful and liveable villages'. This may be based upon the rich culture and appropriate, traditional architecture of the *She* ethnic group, currently celebrated in the museum and evidenced by the iconic *She* bridges and mountain villages, many of which are being lost through urbanisation. A task force of artists and artisans may be assembled to adapt these traditions to modern construction, bringing them out of the museum and revitalising them in the current circumstances of Jingning, so that the county and eco-communities become a model for sustainable, symbiotic and culturally appropriate design—with recognised model villages (as in the case of Baisha). This would not only support ecotourism but also provide employment for local communities and revive some traditional skills. Such a scenario could be supported by Zhejiang universities, including their architectural schools, supplemented by external advisers, standards, guidelines and an accreditation system. It is important to establish and implement such a plan promptly, before the unique natural attractions are damaged by inappropriate, uncontrolled development. Jingning is at the crossroads, both literally and metaphorically.

More accessible villages, such as that located at the base of the mountain near Jingning City, may be revitalised by food enterprises and ecotourism, incorporating modern amenities sensitively integrated with the natural landscape and ecology, cultural artifacts and buildings. In addition, urban communities—where more remote villagers are resettled—could similarly incorporate ecological development features, with the villagers retaining ready access to adjacent hillsides and mountains upon which their agricultural livelihoods were historically based. The various symbiotic development models proposed for Japan rural areas by Takeuchi et al. (1998) could be adapted by and provide an inspiration to Jingning, accompanied by micro-hydro and similar decentralised, sustainable technologies—enabling further opportunities for local employment. As discussed earlier in Chap. 5, alternative energy development involving a 'whole system approach' may not only deliver affordable energy services, but also improve village livelihoods, providing local employment in a gender balanced way and fostering overall economic and human development.

Another scenario for Jingning could involve expansion of the natural and organic food, tea and medicine industries, and their marketing, certification and export to wider markets through green supply chains and services (see Jégou & Joore, 2004). This could even extend to direct supply of food products to the consumer from the producer, rather than using other partners in the supply chain. For example, Odin Holland operates an organic food subscription service direct to consumers, forming strong producer-customer relationships (Manzini & Vezzoli, 2002). In the Jingning case, local producers and their knowledge and resources could form the basis for a wider task force. This could extend to health professionals, as in the case of the Jingning nutritious red rice. The PSS configuration will also require construction of clean processing and packaging facilities, which will require careful planning so they are harmonious with natural ecosystems and landscapes. Much could be gained by learning and applying the principles of industrial ecology, which are especially well-established in Europe (see Tilley, 2003).

The planning and successful implementation of such scenarios and modes will be dependent on increasing the confidence and capabilities of the Jingning *She* people, through training, facilitation and advice for local entrepreneurs—as recognised by the chief of Daxi village (C). For example, one of the most important elements of the OVOP scheme has been human resource development, enabling farmers and producers to gain valuable training from external experts in agro-processing, quality control, packaging and marketing (Wahlin & Natsuda, 2008).

As in the case of Mangjing village, trading of forest property rights and carbon credits may constitute another major opportunity for Jingning to boost rural incomes, employing the methods described in Chap. 4 and including infrastructure improvements within negotiations with afforestation companies.

Baisha Village

Eco-development and construction of tourism and housing facilities is also an important scenario and mode for Baisha village, although this lacks the rich ethnic culture and traditions of Mangjing, Daxi and Jingning. A task force may be drawn from the local tourism operators and contractor, with external experts and training providers drawn from a possible sustainable design advisory panel, and accompanied by standards and accreditation. In the same way that the village building contractor undertakes work for neighbouring villages and similar to the micro-hydro advisory services provided by Daxi village, Baisha may develop expertise, enterprises and diversify its employment base by means of sustainable living and offering design services to other villages.

A related scenario could involve more eco-efficient infrastructure services such as waste management and recycling, sanitation, energy and water. Again, the principles of industrial ecology and symbiosis could be applied, with reference to the models put forward by Takeuchi et al. (1998) and others and the whole system approach to energy services described in Chap. 5. Solid and liquid waste and biomass generated by tourists and agriculture could be treated, reused and recycled, including its conversion to energy—as in the case of Gaviotas, Columbia (Nicholls, 2006).

The PSS mode could include training and accreditation of tourism operators and awareness-raising of tourists themselves. Indeed, Baisha may widen its tourism market by attracting those tourists who are interested in learning about sustainable modes of development.

Baisha village and the associated Dong Keng community, Lin'an, are already seeking to boost incomes by forest carbon sequestration and trade, especially using bamboo, and this provides an opportunity for provision of services and advice to other communities in China – as is already being pursued by the Service Centre for Modern Science and Technology for Forestry, Lin'an.

6.5.2.5 Holistic Consideration and Synergies

Possible connections and synergies among various scenarios and PSS modes can now be explored in a holistic manner by taking a wider perspective, using the zooming out technique described earlier. In addition, finding connections between PSS modes, as explained in Fig. 6.2, enables the actors to share their capabilities, resources and infrastructures, thereby using these in a more efficient and cost-effective manner. Again, further research is required to engage more deeply with the villagers and assist them to express their own ideas.

Mangjing Village

The unique symbiosis of tea growing with biodiversity under the forest canopies may extend to an integrated development strategy for Mangjing. Tea customers and other tourists could be attracted to the mountain village to experience its pure tea, healthy clean air, the ancient *Bulong* heritage and culture, costumes, dance, architecture, and rituals associated with the worship of the 'spirit of tea' and serving of tea (Feng, 2010). Thus, the unique tea product, biodiversity and local heritage could, in combination, provide a number of integrated 'services' (Ayers, 1999). The various actors (such as tea producers, tourism operators and cultural groups) may share their capabilities, resources and infrastructures. For example, inspection of the tea processing facilities provides a unique experience for eco and cultural tourists. Partnerships with external organisations and companies, including PPPs, may inject funds into the village infrastructure, which may serve multiple purposes.

Daxi and Jingning

Various PSS modes such as ecotourism, culture (language, art, built heritage, costumes and dance) and clean, organic food may be interconnected. This could provide an opportunity for the various actors involved with each mode to share their capabilities, resources and infrastructures. For example, infrastructure and cultural facilities may be combined to serve multiple purposes, as in the case of the unique *She* bridge structure within Daxi village. This not only provides a river crossing but also a cultural performance and tourism facility. In combination with nearby cafes serving local produce, and the natural landscape, a unique overall tourism, cultural and culinary experience is provided.

Furthermore, customer markets for organic foods, tea and culture may be a source of tourists, who visit Jingning to appreciate the natural products, artifacts and processes as part of a close customer-provider P-S relationship. This would all assist marketing Jingning to nearby cities such as Shanghai and Hangzhou, in addition to nationally and internationally. This would enable Jingning to tap into global experts in organic and healthy foods, sustainable design and development, and conservation of culture, language and architectural heritage.

Eco-services from the rich natural environment and forests may be integrated with trade in forestry property rights and sequestered carbon.

Baisha

The scenarios and modes of ecotourism, food and appropriate, harmonious facilities and eco-efficient infrastructure may be interconnected. Restaurants and food appreciation, education and cooking facilities may be combined with tourist hotels and adjacent gardens, accompanied by possible hospitality training. Food and tourist solid and liquid waste may be treated, recycled, and generate energy, in addition to compost for food gardens. Baisha may develop expertise and a diversified employment base from sustainable living and design, with the potential to provide services to other villages, as in the case of existing bamboo carbon sequestration services and advice.

6.6 Discussion and Conclusions

The method for community transformation was applied to case studies of remote rural villages in China, with particular reference to services and infrastructure. Table 6.1 presents a summary of the main characteristics of the case study villages and the opportunities for synergies that may arise by use of the method.

Returning to the research questions posed in Sect. 6.2, the findings indicate that such connected-up and holistic approaches may foster partnerships and creative synergies among various actors, which involve sharing of capabilities, resources and infrastructures. In turn, it has been illustrated that this is likely to lead to win-win solutions that result in improved community outcomes with less resource use and less cost, that is, 'do more with less'. Integrated, inclusive and appropriate infrastructure solutions may achieve multiple purposes in a more efficient and effective manner, supporting and underpinning the delivery of products and services by social entrepreneurs and community enterprises. The provision of infrastructure itself may open up export and other opportunities for such entrepreneurs and enterprises, such as in the case of micro-hydro in Jingning autonomous county or provision of tourist facilities in the case of Baisha village, Linan. Thus, the case studies illustrate that synergistic, integrated and inclusive planning of services and infrastructure may play a major role in facilitating stronger communities in rural China and beyond.

Table 6.1 Summary of application of method

Model & processes	Rural communities in China Mangjing (*Bulang*)	Daxi/Jingning (*She*)	Baisha
(a) Understanding the problem & context *(Where are we now?)*	Poverty stricken, although employment by private (Brilliant) tea company has improved livelihoods, incomes	Daxi beautiful mountain setting, natural foods, white tea, red rice	Already underwent transformation from economy based on cutting forest to conservation & cultivation, & achieved recognition as a model eco-village
		Ancient & unique architecture, costumes & language	
	Ancient ethnic group, rich culture	Micro-hydro technology	Tourism accounts for 65 % of GDP, although seasonal, also tea growing. Need for more diversified employment
	Tea plantations exhibit biodiversity	Little industry	
	Uncoordinated tea enterprises	Transport access recently improved	Overdevelopment e.g., large houses, tourist hotels, threaten fragile environment & increase inequity
	Inadequate & uncoordinated infrastructure e.g., electricity, water	Communities are at crucial stage in development to realise opportunities	
	Very remote, poor transport access to markets	Jingning & other mountain villages being closed, people resettled in urban areas	Assistance with sustainable design is sought
(b) Transformation statement/goal *(Where do we want to get to?)*	Grow & compete with neighbouring private tea company, enabling villagers to prosper & control own destiny & own land, natural assets	Investment in hard & soft infrastructure (e.g., education) so people better prepared, increase capabilities.	New transformation goal for households & ecotourism development to be in harmony with natural setting i.e., eco-development, reduce waste via recycling systems, & increase diversification of employment
	Common tea brand, shared facilities & infrastructures	Preserve traditional *She* culture & architecture, ecological development, promote natural foods, tea & medicines, develop beautiful & livable villages	
	Coordinated, shared infrastructure	Develop marketing strategy & brand to attract more visitors, increase markets	
		Jingning village potential to be transformed into ecological tourist village if incorporates modern amenities, infrastructure	

(c) Solution oriented partnerships *(How do we get there?)*	Village council & cooperative, women's group (e.g dance), entrepreneurs	Village council & cooperatives	Village council & cooperatives
		Capabilities & capacity in natural foods & products	Members with capabilities and capacity in holiday houses, tourism. One was acknowledged leader. Also a builder. Forms nucleus of network also drawing upon Dong Keng know-how & external experts
	Knowledge in construction, lighting from methane and waste, and reusing water to create steam for tea drying	Increased management, business & marketing skills required	
		Particular expertise in micro-hydro	
	Brilliant tea company knowledge. Ford Foundation, UNESCO and universities		
(d) Product-service modes *(What do we need to get there?)*	Common tea brand & shared facility & infrastructure such as energy & water reuse	Increased ecotourism, using pristine natural features, clean air, foods & forests, with facilities & architecture	Ecotourism, sustainable & ecological development & construction of tourism & housing facilities
	Culturally appropriate architecture to support ecotourism. Additional cost offset by partnerships with external tourism providers	in harmony with landscape. Based upon rich *She* culture & appropriate, traditional architecture e.g., iconic multi-purpose bridges	Task force from local operators, with external experts, design advisory panels, standards, accreditation.
			Eco-efficient infrastructure services e.g. waste management & recycling, sanitation, energy, water. Apply industrial ecology & symbiosis principles
	Opportunities for trade in forestry property rights and carbon sequestration, with infrastructure improvements being negotiated as part of partnerships with afforestation companies	Natural & organic food, tea, medicine production for new markets & export	
		Green supply chains, eco-industries, clean processing, employing industrial ecology principles	Carbon sequestration expertise in Lin'an provides an opportunity for improved incomes and services to other communities, as is already being pursued
		Opportunities for trade in forestry property rights and carbon sequestration	

(continued)

Table 6.1 (continued)

(e) Holistic considerations	The unique symbiosis of tea growing with biodiversity under the forest canopies may extend to an integrated development strategy for Mangjing	Connect ecotourism, culture, clean organic food scenarios & modes, with forestry eco-services being integrated with trade in forestry property rights and sequestered carbon.	The scenarios & modes of ecotourism, food & appropriate, harmonious facilities & eco-efficient infrastructure may be interconnected
	Tea customers & other tourists could be attracted to the mountain village to experience its pure tea, the clean air, the ancient *Bulong* heritage & culture, costumes, dance, architecture, and rituals associated with serving of tea & the worship of the 'spirit of tea'. Thus, the unique tea product would be a conduit for a number of associated services. Partnerships with external organizations & companies e.g. afforestation companies, including PPPs, may inject funds into the village infrastructure	Restaurants & local food promotion, hospitality & cooking education & gardens combined with hotels	Restaurants & food appreciation, education & cooking facilities may be combined with tourist hotels & adjacent gardens, accompanied by possible hospitality training. Food & tourist solid & liquid waste may be treated & recycled, & generate energy in addition to compost for food gardens
		Food & solid/liquid tourist waste treated & recycled, generate energy & compost for gardens	
			Baisha may develop expertise & diversified employment base from sustainable living & design, & provide services to other villages, as in the case of bamboo carbon sequestration services

Further empirical research is also required to test the indicative ideas put forward, especially to elicit and support the ideas of the stakeholders within the case study villages. More rigorous research is also required to demonstrate, via use of optimisation modelling, measurement and evaluation, whether the approaches outlined in this paper may achieve integrated and synergistic sustainability, having regard to its social, economic, cultural and environmental dimensions.

The next chapter examines one aspect of infrastructure services and systems in more depth, namely, rural ICT service innovation.

References

Alexander, C. (1964). *Notes on the synthesis of form*. Cambridge, MA: Harvard University Press.

Australian Asset Management Collaborative Group. (2011, November 14). *Guide to integrated strategic asset management*. Retrieved December 9, 2013, from http://www.apcc.gov.au/ALLAPCC/AAMCoG%20Guide%20to%20Integrated%20Strategic%20Asset%20Management%202011.pdf

Ayers, R. (1999). *Products as service carriers: Should we kill the messenger – Or send it back?* Tokyo, Japan: Zero Emissions Forum, UN University.

Checkland, P. (2001). Soft systems methodology. In J. Rosenhead & J. Mingers (Eds.), *Rational analysis for a problematic world revisited: Problem structuring methods for complexity, uncertainty and conflict* (2nd ed.). Chichester: Wiley.

Chertow, R. (2000). Industrial symbiosis: Literature and taxonomy. *Annual Review of Energy and Environment, 25*, 313–337. Retrieved December 9, 2013, from http://www.annualreviews.org/doi/pdf/10.1146/annurev.energy.25.1.313

Corning, P. A. (1998). The synergism hypothesis: On the concept of synergy and its role in evolution of complex system. *Journal of Social and Evolutionary Systems, 21*(2). Retrieved February 8, 2011, from http://www.complexsystems.org/publications/synhypo.html

Devisscher, T., & Mont, O. (2008). An analysis of a product-service system in Bolivia: Coffee in Yungas. *International Journal Innovation and Sustainable Development, 3*(3/4), 262–284.

Dewey, J. (1910). *How we think*. New York: Dover.

Feng, J. (2010, December 7). Transforming the rural tourism trade. *Shanghai Daily Hangzhou Special*. Retrieved http://www.hicenter.cn/columns_detail.asp?id=434942

Hung, P. Y. (2012). *On a frontier tea road – Mountains of green gold: Tea production, land use politics, and the Bulang People in Southwest China*. Research Statement. Retrieved October 6, 2012, from http://poyihung.wordpress.com/research-statement/

Jackson, M. C. (2006). Creative holism: A critical systems approach to complex problem situations. *Systems Research and Behavioural Science, 23*, 647–657.

Jégou, F., & Joore, P. (2004). *Food delivery solutions. Cases of solution oriented partnership*. Bedford, UK: Cranfield University.

Joham, C., Metcalfe, M., & Sastrowardoyo, S. (2009). Project conceptualisation using pragmatic method. *International Journal of Project Management, 27*, 787–794.

Krucken, L., & Meroni, A. (2006). Building stakeholder networks to develop and deliver product-service-systems: Practical experiences on elaborating pro-active materials for communication. *Journal of Cleaner Production, 14*, 1502–1508.

Liang, L. (2010, April 7). The tea forests of Yunnan. *Our World*. United Nations University. Retrieved January 6, 2014, from http://ourworld.unu.edu/en/the-tea-forests-of-yunnan

Manzini, E., Collina, L., & Evans, S. (2004). *Solution oriented partnership: How to design industrialised sustainable solutions*. Bedford, UK: Cranfield University.

Manzini, E., & Vezzoli, C. (2002). *Product-service systems and sustainability: Opportunities for sustainable solutions*. Paris: UN Environment Programme (UNEP), Division of Technology, Industry and Economics, Production and Consumption Branch.

Metcalfe, M. (2007). Problem conceptualization using idea networks. *Systematic Practice and Action Research, 20*, 141–150.

Mont, O. (2004). *Product-service systems: Panacea or myth?* (p. 259). Lund, Sweden: IIIEE, Lund University.

Morelli, N. (2006). Developing new product-service systems (PSS): Methodologies and operational tools. *Journal of Cleaner Production, 14*, 1495–1501.

Ness, D. (2008). Sustainable urban infrastructure in China: Towards a Factor 10 improvement in resource efficiency through integrated infrastructure systems. *The International Journal of Sustainable Development and World Ecology, 15*(4), 288–301.

Ness, D., & Xing, K. (2010). A theoretical construct of a product-service systems model for low social-economic-status communities. In *Proceedings of INFORMS service science conference*, Taipei, Taiwan.

Nicholls, R. (2006). Toward an ecology of peace, World and I: Innovative approaches to peace. *Winter*, pp. 71–79.

Office for Infrastructure Development. (2005). Strategic infrastructure plan for South Australia 2005/6 to 2014/15. SA Government.

Partidário, P., Lambert, J., & Evans, S. (2007). Building more sustainable solutions in production-consumption systems: The case for people with reduced access. *Journal of Cleaner Production, 15*, 513–524.

Pawar, K. S., Beltagui, A., & Riedel, J. C. K. H. (2009). The PSO triangle: Designing product, service, and organisation to create value. *International Journal of Operations and Production Management, 29*(5), 468–493.

Ramirez, R. (1999). Value co-production: Intellectual origins and implications for practice and research. *Strategic Management Journal, 20*, 49–65.

Ritthoff, M., Holger, R., & Liedtke, C. (2002). *Calculating MIPS: Resource productivity of products and services.* Wuppertal, Germany: Institute of Climate, Environment. and Energy.

Schmidt-Bleek, F. (2000). *The factor 10/MIPS-concept bridging ecological, economic, and social dimensions with sustainability indicators.* The Factor 10 Institute. Retrieved December 9, 2013, from http://www.factor10-institute.org/files/MIPS.pdf

Sinclair, J. (1995). *Collins COBUILD English Dictionary*. London: HarperCollins.

Takeuchi, K., Namiki, Y., & Tanaka, H. (1998). Designing eco-villages for revitalising Japanese rural areas. *Ecological Engineering, 11*, 177–197.

Tilley, D. R. (2003). Industrial ecology and ecological engineering: Opportunities for symbiosis. *Journal of Industrial Ecology, 7*(2), 13–32.

UNESCAP. (2011). *Are we building competitive and liveable cities? Guidelines for developing eco-efficient and socially inclusive infrastructure.* Environment and Development Division, United Nations Publication, Bangkok. Retrieved November 23, 2011, from http://www.unescap.org/esd/environment/infra/documents/Guidelines.pdf

Vargo, S., Maglio, P. P., & Akaka, M. A. (2008). On value and value co-creation: A service systems and service logic perspective. *European Journal of Management, 26*, 145–152.

Von Weizsäcker, E., Hargroves, K., Smith, M., Desha, C., & Stasinopoulos, P. (2009). *Factor 5: Transforming the global economy through 80 % increase in resource productivity.* London/Germany: Earthscan/Droemer.

Wahlin, W., & Natsuda, K. (2008, January 15). One village one product: Japan's rural entrepreneurial scheme goes abroad. *Japan Inc Magazine*, 75. Retrieved November 12, 2011, from http://www.japaninc.com/mgz_jan-feb_2008_one-village-one-product

WBCSD. (2000). *Eco-efficiency: Creating more value with less impact.* Conches-Geneva, Switzerland: World Business Council for Sustainable Development.

World Bank. (2012). *World development report 2013: Jobs, International Bank for Reconstruction and Development*, Washington. Retrieved October 5, 2012, from http://siteresources.worldbank.org/EXTNWDR2013/Resources/8258024-1320950747192/8260293-1322665883147/Overview.pdf

Xing, K., Ness, D., & Lin, F. (2013). A service innovation model for synergistic community transformation: Integrated application of systems theory and product service systems. *Journal of Cleaner Production, 43*, 93–102.

Chapter 7
Key Characteristics of Rural ICT Service Innovation: A Case Analysis of ICT-Enabled Rural Financial Services in China

7.1 Introduction

Information and communication technologies (ICTs) and ICT infrastructures can benefit rural small to medium enterprises (RSMEs), and rural communities in general, in many ways. For farmers and cooperatives, ICT devices, applications and infrastructures facilitate more efficient and consistent communications with expanded networks that support improved information and services for crop, business, and resource management, new knowledge and techniques for farming and business operations, financial and social capital, certification, and market access (Ratsifandrihamanana, 2012). Cases of new programs and solutions are reported in a number of recent studies and projects on rural ICTs for bridging information and service gaps between rural and urban communities, such as an internet kiosk for sugarcane farmers in India (Gollakota, 2008), mobile agricultural services for farmers in China and Uganda (Kiiza & Pederson, 2012; Mackie, 2010), mobile banking with microfinance in Kenya (Karugu & Mwendwa, 2007) and wireless IP access in rural Bhutan (Tobgyl, n.d.). Although some achievements have been observed, concrete evidence of effective rural ICT services with successful business models is still incomplete. The main challenge is to introduce appropriate ICT assets and related services that are available and affordable for farmers and RSMEs to meet their needs.

The importance and need for collaborative development of ICT-enabled services are emphasised in many existing rural ICT programs and cases. However, there is a lack of a clear and systematic analysis on the key characteristics of ICT applications in rural services to underpin and enable a methodical process for innovative and synergistic social-technological solutions. This chapter explores and identifies the key factors and likely pitfalls in applying ICT solutions to facilitate rural services and enable effective collaboration for ICT-based service innovation to support people and businesses in rural communities. To address these two tasks, we provide a

Y. Zhu et al., *Transforming Rural Communities in China and Beyond*,
DOI 10.1007/978-3-319-11319-7_7

review of the literature on development and limitations of some examples of extant ICT-based rural services. A case study on ICT-based rural financial services in China is then presented in detail. Based on the literature review and the case analysis, the next section discusses some major elements in ICT-based rural service innovation which can be adopted to further construct a systematic model for rural ICT development. The paper concludes by summarising the key findings from the research and what needs to be further developed.

7.2 ICT-Based Services for Rural Communities: Development and Limitations

Basically, ICTs are referred to as a range of technological applications that can facilitate information processing and exchange in electronic forms. Key elements to shape and deliver ICT solutions include ICT infrastructures (e.g., telephone network, broadband network, mobile network, etc.), ICT appliances (e.g., personal computers, mobile phones, wireless PDAs, satellite dishes, software applications, etc.), ICT services (e.g., e-money transfer, e-learning, e-health, web portals, etc.), and their integration. According to Dymond et al. (2010), ICTs for rural communities and businesses in developing countries, to have improved access to information and to meet other developmental needs, are often based on the objective of 'universal access'.

7.2.1 Universal Access for Rural ICTs

Much of the focus of universal access is on developing available, accessible, and affordable technological solutions and business models for communities in certain rural/remote regions to expand, or obtain for the first time, access to the internet, telecommunication networks, or other ICT facilities, appliances and resources from home, community, or public access points (Black & Atkinson, 2007). Availability and accessibility of computers and telecommunication appliances are also important for rural ICT access. As demonstrated in a study by Brouwer and Brito (2012) on the diffusion patterns of mobile phone use in Mozambique, the digital divide—such as in mobile phone access—is in fact underlined by socio-economic and gender inequalities. Their study also highlights that lack of rural infrastructure support and shortage of reasonable business models are two major barriers that greatly restrict ICT access of communities in poor regions. While family or community-based sharing can be a viable way to reduce digital divide in rural areas of developing countries (James, 2011), involvement from the public sector, such as local governments, public administrations, and nongovernmental agencies, in partnership with private corporations, plays an increasingly active role in developing novel business and operating models for affordable access to ICT infrastructures and appliances. Examples of such public-private multi-sectoral collaboration initiatives to achieve universal access include the community information centres established in

Bhutan and Thailand that provide easy-to-access computers and ICT equipment (NECTC, 2003; Tengtrakul & Peha, 2011; Tobgyl, n.d.;), the low-cost telecommunication technologies and devices for rural access in Tanzania (Rugonzibwa, 2012), the Grameen Village Phones project in Bangladesh (Enejo, 2009; Futch & McIntosh, 2009) and the n-Lounge model for rural wireless data and voice communication facilities in India (Rao, 2007).

Meanwhile, there is also an emphasis on enhancing the coverage of access for everyone in a rural community to obtain particular services through novel use of ICT appliances and applications as a critical aspect of universal access. For example, Nongxintong, a farming information service, introduced by China Mobile in rural China delivers information and news to farmers via the mobile network and mobile phone handsets, which increases farmers' access to the markets, customers and farming information and greatly improves efficiency and productivity (Mackie, 2010). ICT-based market information provided through local FM radio stations, mobile phones, and internet facilities to smallholder farmers in Uganda is another successful case of using ICTs as media for access to essential services that have led to improved seeding and agricultural productivity (Kiiza & Pederson, 2012). Similarly, using cloud computing technology to enable rice growers and small merchants in India to gain better price and market information for improved logistics planning (Tsao et al., 2010) is a typical case of an ICT-enabled rural information service. The review conducted by Essegbey and Frempong (2011) on mobile telephones in Ghana also highlights the notion that the use of ICT tools (such as mobile phones) for service innovation requires a more holistic and participatory model, which involves effective collaboration among public and private stakeholders and impetus from both the demand and the supply sides.

7.2.2 *ICT-Based Rural Service Innovations: Two Examples*

As reported in some extant projects and studies of ICT for rural development (see Bakhshizadeh, Hosseinpour, & Pahlevanzadeh, 2011; Black & Atkinson, 2007; Mayeku, Kilwake, & Bertarelli, 2010; Rao, 2007), the most important role of ICTs is to promote viable and efficient accesses to essential services (e.g., rural education and training, health, local and wider community networking, market and business information, entrepreneurship development, etc.) that enhance natural, human, financial, physical and social capital for livelihood development of the rural poor (Chapman & Slaymaker, 2002). While innovative ICT devices, such as VillageCell (Anand, Pejovic, Johnson, & Belding, 2012) and GSM Phone Booth (Enejo, 2009), can improve universal service access, it is more important to have rural ICT applications—fit for local characteristics and conducive for rural service innovations—that can contribute to investment in local capacity building, locally-generated investment, and active local control (Bakhshizadeh et al., 2011). Two typical examples that present innovative ICT practice for improving rural services are M-PESA's mobile money service in Kenya, and Indiagriline's information services in Tamilnadu, India.

Table 7.1 M-PESA partnership network

Partner	Role	Capability
Vodaphone	Mobile network provider	Mobile banking technology and business investment
Dept for Int'l Development (DFID) (UK)	Arm of the British government	Funding for costs of the pilot project
Safaricom Kenya	Leading mobile phone operator in Kenya	Communication services and devices
Faulu Kenya	Microfinance organisation	Loan services for low-income earners and operations networks
Commercial Bank of Africa	Private bank	Personal banking services
Microsave	NGO funded by DFID	Microfinance expertise and industry experience

7.2.2.1 M-PESA

The M-PESA project is a typical example of public-private collaboration in providing improved access to basic financial services in rural Kenya through an innovative use of ICTs (Karugu & Mwendwa, 2007). The project enables mobile banking (e.g., cash withdrawal, cash deposits, and bill payments) and funds transfer (between individuals as well as between individuals and businesses) by using mobile phones and registered local retail outlets. Initiated by Vodafone, it involves a number of partners that provide support via funding, business and technical capabilities, and service products, which are summarised in Table 7.1.

The pilot of the M-PESA system was launched in Kenya in late 2005. It allows rural customers to use mobile phones, with a specially designed SIM card and software, as a debit card and to make electronic transactions through SMS. Clients of the system hold special M-PESA mobile phone accounts. Cash transactions can be done at Safaricom shop outlets by transferring credit from their phones to the shop in exchange for cash, or vice versa. Such mobile phone-enabled financial services make it relatively easier for farmers to conduct personal banking and loan payments, without the need for carrying cash, extended travel, and queuing. However, the transaction services at the local mobile phone shops do incur some extra costs (up to US$5.71 for US$500 money transfer).

Although overall the M-PESA pilot represents an effective ITC-based service innovation, the new system also faces some major challenges. Firstly, the low income farmers are often under-educated and unfamiliar with both banking procedures and sophisticated use of mobile phone and text messaging functions for account management. Additional training is needed for both farmers and customer service staff at local retail shops for using the M-PESA system. Meanwhile, the limited coverage of mobile network and mobile communication services in rural Kenya, as well as potential regulatory implications of mobile banking, also hinder the expansion of the M-PESA implementation.

Table 7.2 Indiagriline project partners

Partners	Capability
EID Parry, Sugar manufacturer	Providing internet, computer kiosk, and web portal for accessing farming-related technical and business information, ERP access, and other generation information
ICICI Bank, Indian Bank	Banking and loan services
Hewlett Packard	Providing 'all-in-one' devices for rural environment
NIIT	Computer literacy program
Local universities (e.g., Tamilnadu Uni. for VET and Animal Sciences, Tamilnadu Agri. Uni.)	Content development and consultancy for farmers
Tata Consultancy	Adult education software for literacy development
UNESCO	Funding for two women entrepreneurs
Local Community	Service users and providers, kiosk franchisees, and entrepreneurship

7.2.2.2 EID Parry's Indiagriline and Computer Kiosk

Established in 1788, EID Parry is a firm that processes and markets cane sugar from Tamilnadu, India. It receives supplies from 10,000 sugar cane farmers in 100 villages (Gollakota, 2008). From 2000, the firm developed ICT initiatives to provide local farmers improved access to information on farming practices, farm inputs, farm management, market, as well as education, health and other services via rural computer kiosks (Parry's Corners) for the physical connection, and the Indiagriline web portal for the virtual connection. Facilitated by EID Parry, this rural ICT project also involved some key partners from public and business sectors for content development, value adding service delivery, and funding and resource provision, as listed in Table 7.2.

The implementation of computer kiosks was largely through independent franchises and partnerships between franchisees from local villages and a partner bank, complemented by some company owned and operated kiosks. Meanwhile, the web portal was set up and maintained by EID Parry to provide local sugar cane farmers access to farming and Enterprise Resource Planning (ERP) information to manage their produce. While the outcomes of the two initiatives demonstrated some direct benefits for villagers in helping local jobs, local incomes, and community awareness and responsiveness, the ICT resources were not actively used by the wider community as initially expected. It was found that computer-related applications alone (e.g., browsing, chatting, emails, etc.) did not generate much income to sustain the kiosk operations. Rather, innovative use of the ICT facilities by some computer kiosk owners to provide ancillary services (e.g., renting out computers for education, providing exam results, astrology, matchmaking, etc.) and even to sell unrelated goods (e.g., fertiliser, farm products, ice cream, etc.) to walk-in clients proved more attractive to the farmers. As reflected in EID Parry's rural ICT project, enabling access to ICTs as a technical solution for information service alone has limited effectiveness. It is more important to have the ICT access integrated with the development of social capital and entrepreneurship in rural communities.

7.2.3 Characteristics of Rural ICT Services and Current Limitations

It has been well understood that access to information and communication infrastructure, appliances, and services are fundamental for rural development. However, concrete evidence of effectiveness of rural ICT implementation is still incomplete. Many extant studies and projects on rural ICTs are still at the pilot stage and, as reflected in some examples reviewed, the major focus of low-cost ICT access is on technological means to reduce digital divides and to meet development objectives. Although such effort can bring about some positive outcomes for the rural communities affected, the impact will be limited largely by its narrow scope. The development and implementation of new technologies and technological solutions need to be adapted to the conditions of rural and remote regions in developing countries for poor and less/un-skilled users. In effect, rural ICTs are mainly of a social-technological nature. As highlighted by some researchers (e.g., Bakhshizadeh et al., 2011; Chapman & Slaymaker, 2002; Rao, 2007), those relatively successful projects are more effective in providing total and holistic solutions (technological, institutional, and entrepreneurial) to address the needs of the rural poor. In particular, they enable novel use of ICT resources as the medium for service innovations and service access to improve livelihood assets (physical, natural, financial, social, and human capital) of rural communities.

Currently, rural ICT service development and implementation largely relies upon imitating extant examples or learning from best practices. Despite various ICT projects for rural development, the social, economic, cultural, political, and technological characteristics of rural ICTs and ICT-based services differ greatly from case to case. Guidance for proper design and deployment of ICT solutions is not always easy to access and follow, as service innovations are highly contextualised. In contrast to many case studies presented on rural ICTs and their achievements, discussions on generic methodologies for rural ICT development and implementation are rare in current literature. Although there have been some comprehensive reviews of common features and paradigms of ICT-based interventions to inform practice and policy-making for rural development (e.g., Barrett & Slavova, 2011; Caspary & O'Connor, 2003; Chapman & Slaymaker, 2002), to date ICT-based rural service innovations are based more on empirical experience than on following a methodical development process. In this sense, a major limitation in rural ICT applications is the lack of a clear and systematic model to guide development of collaborative partnership and systems solutions for ICT-based service innovation. As an attempt to address this knowledge gap, this chapter seeks to use an in-detail analysis of an ICT-based rural financial service case in China to identify and discuss some key factors and paradigms for facilitating collaborative solution-oriented rural ICT service innovation.

7.3 A Case Study: ICT-Based Rural Financial Service Innovation in Lishui, China

In this section, a case of rural financial services for local communities and small businesses in Lishui, Zhejiang Province of China is analysed. The information used in the case study is mainly based on interviews with local banking and financial service providers, local government officials, and local community members (including villagers, community leaders, and family-based cooperatives), as well as collated from some published resources (e.g., Chen, 2012; Chen & Yao, 2009; Gu & Wang, 2012; Shi, Lin, & Zhou, 2011; Wang & Gu, 2012; Zhang, 2012). Although some data and information may not be precise, they depict the general profile and features of rural financial service development and issues in the region.

7.3.1 Understanding the Context

Located in the southwest of Zhejiang province of China, Lishui prefecture is 17,300 km^2 with nearly 90 % of the region dominated by mountainous landscapes. Much of its 2.1 million population lives in rural areas or mountain villages. While the overall economic development and average income per capita are behind other areas in Zhejiang, Lishui region has a very high forest coverage (approx. 70 %), rich forestry resources and products (e.g., timber, bamboo, tea, tree nuts, etc.), as well as ecological and cultural diversities (e.g., it has the largest settlement of unique ethnic *She* minority groups in Southeast China). Meanwhile, 80 % of the 1.4 million hectares of woodlands in Lishui are allocated to rural households, and the region's economic profile is defined by family-based small forest-product and farmhouse-tourism cooperatives.

7.3.2 Rural Finances in Lishui and Challenges

The unique natural and cultural resources in Lishui provide great potential for eco-tourism and agricultural industries. However, the development of the rural communities in the region, including those local small cooperatives and family businesses, is largely constrained by the effectiveness and efficiency of rural finances. To the villagers and small rural enterprises in Lishui, rural and agricultural subsidies, forest right trading and mortgage, and microloans are three major forms of financial resources that are essential for supporting their livelihood, production, and business operations. Like many similar rural communities in China, there are some major difficulties for local villagers and agri-product cooperatives in Lishui in getting easy and consistent access to these financial resources and services.

- *Rural and agricultural subsidies*: Currently, there are many kinds of subsidies provided by the national and local governments to benefit rural agricultural communities. However, most of these subsidies are monthly payments in small amounts. More than 40 small villages and townships in Lishui still do not have any branch outlet offering over-the-counter banking services. Villagers need to make frequent travels to neighbouring town centres to access banking services for cash payments and managing their bank accounts. For some senior villagers, the cost of travelling to a nearby county to obtain their Elderly Subsidy could be as much as a half of the payment itself, in addition to many other inconveniences. On the other hand, to provide direct over-the-counter banking services in villages will require about 250,000 RMB initial investment and ¥125,000 RMB per year operating cost from banks. Although ATMs have been implemented in a number of townships, they also have problems such as being expensive to operate (200,000 RMB/machine and 50,000 RMB/year to run), having limited functions, being difficult to maintain, difficult to use, and being a security concern.
- *Forest rights trading and mortgage*: In Lishui, each household in the rural area has been allocated a certain amount of woodland (approximately 0.6–1 ha/person). Although the households do not possess woodlands and forests, they can manage and trade the use right of those forestry properties. As a way to turn assets to capital and 'leaves to notes', farmers can use their forest rights to obtain mortgage loans from banks or other financial institutions. For many rural households, this is the only viable way to get substantial capital resources to support expansion of agri-production or to start new businesses. However, offering forest right mortgage and mortgage insurance services is often challenged by lack of efficient information sharing among forest farmers, forest right trading and management agencies, as well as banks regarding forest asset valuation, asset management, and mortgage risk assessment and management. An effective e-platform needs to be established to overcome geographical distances and differences to facilitate efficient forest right exchange, assessment, and registry management.
- *Microloans*: As part of China's rural financial service reforms, farmers and family-based cooperatives in Lishui can use their personal credit to borrow money from local banks to support their production and business activities, without the need to use their houses or other properties as collateral. Farmers now can use and reuse their credit to borrow loans up to ¥50,000 RMB each cycle. As there are several banks providing microloan services in rural areas, farmers often borrow money from different lenders. For local banks and lending institutions, there is a pressing need to establish a credit information system platform where data and information concerning their rural clients can be properly stored and shared for more effective and efficient credit rating and management. Meanwhile, farmers and small cooperatives also need access to such a platform to manage their borrowing credit, to access information about available financial products, as well as to apply for and repay their loans.

7.3.3 Stakeholders for Financial Services in Lishui

The rural finance reform in Lishui began in 2007. It has aimed to establish a special rural finance model that integrates personal banking, a rural credit system for micro-loans, and forest right mortgage services. In such a model, heterogeneous stakeholders that cooperate to produce and deliver those financial services include Lishui's local rural community (including farmers and small rural businesses), financial service institutions (including banks and forest right trading agencies), as well as local governments and government agencies.

As customers of rural financial services, local farmers and small family-based businesses can also participate in the service co-creation and delivery. While they can use their forest rights for mortgages, their capability and entrepreneurship in using the loans for managing and utilising forestry resources for growing forest and other agricultural products, undertaking agri-production, as well as engaging in eco-tourism business (e.g., *Nong Jia Le,* or Farmhouse Stay, tourism) are critical for the sustainability of rural financial services.

Among the participating banking and financial institutions, People's Bank of China (PBC), as the Central Bank, is instrumental in developing, implementing, and regulating the rules, mechanisms, and credit information system for the rural finance model in Lishui. Commercial banks, such as Postal Savings Bank, Agricultural Bank of China, and Construction Bank of China, provide a variety of personal and business banking and credit products, credit rating and risk management, as well as investments in setting up banking facilities and hardware to rural communities through their branch offices and ATMs in counties and townships. With their branches at the township or even village level, Rural Credit Cooperatives (RCCs) have the grass-roots reach in rural areas and operate as the main legal financial institution (other than banks) providing microfinance to serve the credit needs of rural enterprises and individuals that cannot be met by other commercial banks. For forest rights trading and mortgage loans, the Forestry Exchange Centre (FEC) and Forestry Resource Assessment Agency (FRAA) are two key frontline service providers. The role of FEC is to provide platforms and manage networks for collecting and publishing forest property rights transfer information as well as organising forest rights trading activities. FRAA offers valuation services that are critical for forestry asset auction, mortgage, leasing and liquidation.

Meanwhile, local governments and some government agencies are also involved in facilitating this rural finance model. Forest Right Management Centre (FRMC) and Forest Resource Purchase and Storage Centre (FRPSC) are two government agencies that support forest farmers, cooperatives, as well as forest rights mortgage issuers with key services such as property rights registry management, mortgage insurance, and stockpile and disposal of secured forestry resources. Lishui City and other levels of government are able to provide a policy framework and funding to support low-interest mortgage loans and forestry resource insurance for farmers, as well as other related infrastructure and capacity development projects. Table 7.3

Table 7.3 Key partners of the Lishui rural financial service initiative

Partners		Capabilities and resources
Rural Community	Farmers and small family-based businesses	Use right of allocated woodland, forest products, ecotourism resources, agri-production cooperatives, and entrepreneurship
	Local grocery retailers	Shops, access to/for local customers, basic accounting and money flow management skills, and entrepreneurship
Financial Service Institutions	People's Bank of China	Expertise and networks of branches (at the city or county level) for regulating financial market, financial services and relevant financial business operations
	Commercial banks (e.g., Postal Savings Bank, Agricultural Bank, etc.)	General banking and loan services, investment, financial expertise and operational skills, networks of branches (at the county or town level), credit rating and credit information management
	Rural Credit Cooperative	Local access and knowledge, network of rural outlets (at the village level), localised financial services for small deposits and small loans, credit rating and credit risk management
	Forestry Resources Assessment Agencies	Forestry assets valuation, auction, mortgage, leasing, liquidation valuation services
	Forestry Exchange Centre	Platform, network, and expertise for forest rights and forestry resource trading
Government Agencies	Forest Rights Management Centre	Issue forest rights certificates, records and right transfer management, dispute mediation and arbitration, forest right mortgage registration
	Forestry Resources Purchase and Storage Centre	Acting as guarantor for forest right mortgage, risk management, purchase back mortgaged forest right when bad credit happens
	Local governments	Administrative resources and support, infrastructures, local knowledge, funding support

presents a summary of the key functions, capability and resources, and limitations of some major players that form the collaborative stakeholder network for the endeavour.

7.3.4 Common Goal and ICT-Based Financial Service Platform Innovations

Personal banking, microfinance, and forest rights mortgage are three main financial services that are required by the farmers and small to medium family-based cooperatives in Lishui to meet the needs of living and production, similar to their urban counterparts. Governments at all levels in Lishui support the new rural finance model as a way to help rural empowerment, poverty alleviation, and sustainable development objectives. For the financial institutions, providing rural communities with better access to microloans, forest right mortgage and other financial services also represents benefits for business opportunities as well as corporate social responsibility.

However, most of the financial service providers do not have a physical presence via facilities or staff in villages or small townships. Also, they often operate in isolation and in competition. As discussed in the previous sections, the main challenge for current rural financial service offerings and finance reform in Lishui is not with the design of rural finance products. Rather, it is more related to the way in which they are delivered and with the platforms to facilitate collaboration and resource sharing among the stakeholders for the efficiency of service provision and access. Therefore, based on the interests of different stakeholders and current obstacles, the common goal for the collaboration can be stated as:

> Developing convenient, efficient, and cost-effective platforms to provide rural communities and small businesses in Lishui with improved, reliable and low-cost local access to financial information, banking, and capital services, as well as for integrated information management and sharing among financial service institutions.

With a common goal for service improvement, the following two ICT-based solutions are identified to enable the platforms for facilitating the efficiency and effectiveness of the three rural financial services:

- Point of Sale (POS) system for personal banking service at local grocery stores
- e-platforms for credit loan and forest property right exchange and mortgage services.

7.3.4.1 POS System for Personal Banking

For villagers and small rural businesses in Lishui, small cash withdrawals and deposits have always been a problem due to the limited presence of bank branches in the rural area. Although ATMs have been implemented in a number of townships, they are costly to maintain and have limited banking service functions. Moreover, most villagers, particularly the elderly, have limited education and account management skills and find ATMs inconvenient and very challenging to use. For these rural customers, self-service using online banking through web browsers or mobile banking through mobile phones is an even less feasible option. While ICT infrastructures and appliances are capable of enabling electronic transactions and efficient banking services decoupled from bank counters, as technological solutions they are often limited by social obstacles.

By collectively examining the interests, capabilities and limitations of the stakeholders and ICT alternatives for personal banking, it is identified that implementing POS systems at local convenience stores presents the best socio-technological solution model. A POS system is an embodiment of ICT hardware (e.g., computer, monitor and display screens, cash drawer, card reader, receipt printer, pin pad, etc.), application software for debit card transactions and customer account information processing, and network infrastructure (i.e., internet connection). In this model, some commercial banks (e.g., Agricultural Bank and Postal Savings Bank) will issue their rural customers debit or cash cards and provide qualified local store owners (e.g., those with business skills, good reputations, and who have passed a background check) with a POS machine, a safe, a counterfeit detector, and necessary book-keeping and banking service skill training. While POS systems are used as a cash register for business transactions when customers purchase goods from the stores, at the same time they can also be used for cash withdrawals and deposits, bank account enquiries, and fund transfers of small amounts. Currently, a customer can use POS systems at local stores to withdraw cash up to three times a week with maximum 1,000 RMB every withdrawal.

For local villagers and small businesses, a POS system and the involvement of store owners (who themselves are often local residents) can provide easier and more reliable access to basic banking services at the village level. For banks, with the POS system and ICT, convenience stores effectively play the role of bank counters with only 2,000 RMB set-up cost per POS banking system and 1,000 RMB per year operation cost. As banks only need to process and manage transaction data and client account information, logistics implications are reduced. Participating store owners, acting as the agent for both customers and banks in handling cash transactions, also benefit from assistance from banks and subsidies from local governments, as well as extra sales of goods from increased visits of local customers for banking services. In addition, although every banking transaction record is processed by POS system and transmitted to the related banks, the store owners often keep the received cash incomes and deposits from sales and banking services in their safes locally. As a result, they see increased cash flows and use them as a buffer to meet not only cash withdrawal demands, but also their own business needs. This effectively reduces their operational costs and minimises travel to bank branches. Some entrepreneurial store owners have used this opportunity to improve their own business efficiency and business portfolios.

7.3.4.2 E-platforms for Rural Credit Loan and Forest Right Mortgage Services

Information asymmetry and the lack of an established credit system are the major barriers that hinder many financial institutions from developing and delivering credit loan services in rural areas. Since 2009, the Municipal Government of Lishui has committed significant amounts of funding and administrative resources to collate and compile the rural credit rating information that so far has covered all the villages and 92 % of the rural households in Lishui (Zhang, 2012). On this basis, the PBC Lishui branch has further developed the Rural Household Credit Information System that manages the database of credit and loan borrowing histories as well as credit assessment records of over 380,000 rural households. Such information can be used as the basis for issuing a Household Credit Loan Certificate, which is not only critical for the farmers and small businesses to apply for microloans, but is also important for the assessment and approval of forest rights mortgages. The system and its database provides a unified information service platform for financial institutions to access and check the credit-relevant information of their rural clients, such as how many houses they own, how much forest land they have, whether they have any criminal record, and their credit rating levels. With such a platform, all financial institutions operating in Lishui can now have cost-free information sharing as well as equal opportunities to compete in the rural market. Commercial banks that do not have their own retail network in rural areas can assess credit and approve loans by using the information system to acquire the required credit data, without the need to conduct a house-by-house survey or to set up their service outlets in villages. Both the cost and the time of credit loan service are thus greatly reduced, making the microfinance more accessible and efficient.

For forest farmers and small forestry product cooperatives, using their forest property rights for mortgage is an important source of major capital for their business operations. However, they often have difficulties accessing information about different mortgage products available. Also, due to very limited presence of relevant agencies in the rural area, delivering and accessing key services, such as forest resource asset evaluation, forest property registration, credit assessment and loan approval, and forest right exchange, can also be time and resource consuming. Essentially, forest right mortgage services are dependent on interconnected roles and functions of local farmers, FRAA, FRMC, FEC, FRPSC, RCCs, banks, as well as local governments. Lishui is among the first regions in China to develop ICT solutions to provide an effective electronic platform, or e-platform, for efficient collaborations among the stakeholders and digitalisation of the service provision and access.

As illustrated in Fig. 7.1, the e-platform consists of interconnected databases, an information management system, and a web portal for online forest rights exchange and forestry credit loan products, managed and operated by different stakeholders, with the access embedded with and facilitated by local operations of RCC branches in villages.

- The Regional Forestry Resource and Asset Database contains digitised maps that store the data and information of rural households' forestry asset profiles, such as types and conditions of trees, geographical conditions, boundaries and sizes of properties, etc., providing support for asset valuation and risk assessment.
- The Household Forest Property Rights Information Database records forest rights status, forest land property registration, and histories of forest rights certification, changes, and mortgages for each registered rural household. The database is maintained by FRMC and can be accessed by all the stakeholders to inform forest rights management and exchange.
- The Forestry Asset and Property Rights Information Management System was established to integrate and manage the real time access to the two databases that are maintained by separate forestry organisations. Such a system provides a digital platform for forest rights management and financial service providers

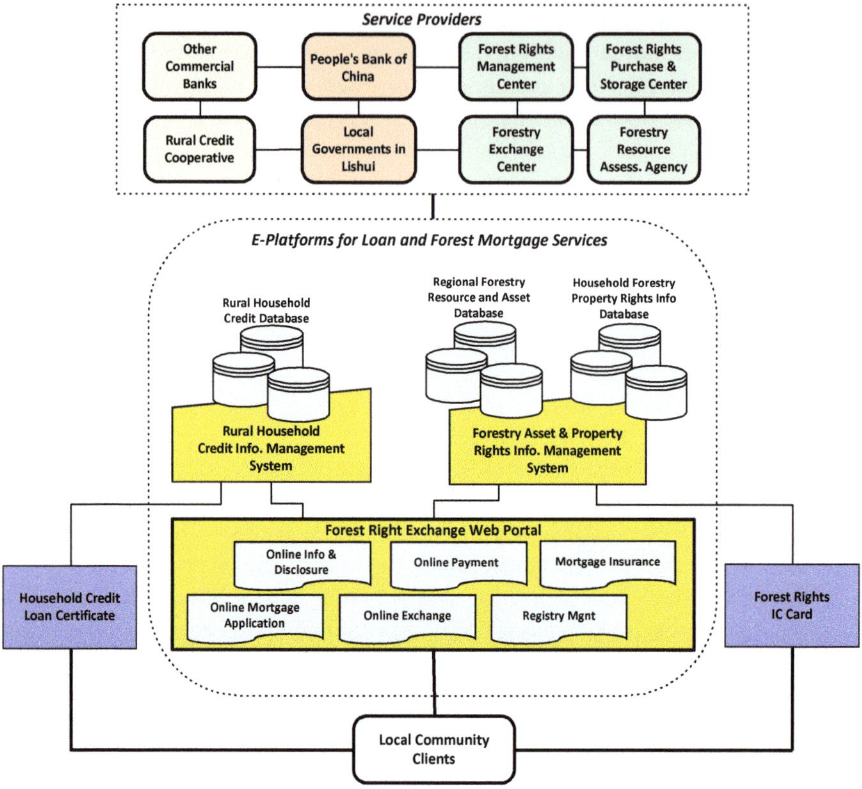

Fig. 7.1 E-platforms and capital services for rural communities in Lishui

(as listed in Table 7.3) to communicate and share information efficiently. In association with this information management system, an IC Card (also known as Smart Card) storing personal and forestry asset information can be issued for each local household and forest product cooperative to access the platform. As the local access point for this information system is at RCC branch offices in villages, using the IC Card, together with a credit loan certificate issued by the RCC, farmers can easily have their asset information reviewed and mortgage loan application processed locally, without the need for paperwork and travel.

- The Forestry Property Rights Exchange web portal is a common trading platform. Through the web portal, both forest farmers and forestry resource businesses can register for online access of policy and trading information disclosure, property rights exchange bidding, and payment transaction services, as well as different forest rights mortgage products, forest asset valuation, and mortgage insurance services. By linking with the Forestry Resource and Asset Information Management System and the databases, the web portal is also an access point for accepting and processing forest rights mortgage applications online. This is especially convenient for those young farmers and forestry resource entrepreneurs who are computer literate and familiar with the use of the internet and web-based applications.

7.3.5 Outcomes

As described above, the focus of financial services for the rural communities in Lishui is to enable improved local access to personal banking, forest rights mortgages, and microloans, with the support of ICT networks and ICT service solutions. POS systems at village convenience stores and databases for both forestry resource and asset information and also forest rights management are developed in forms of ICT-based service innovation to simplify and streamline the rural financial service provisions. So far, the results have been quite effective. By May 2012, 2,114 of 2,852 local villages in Lishui had implemented POS systems for personal banking. In total over 370,000 cash transactions (amounting to 73 million RMB), using the POS system service at local stores, have been recorded (Gu & Wang, 2012). At the same time, databases have been established to record the digital profiles of 14,000 km^2 of forestland and 485,000 forest rights registrations throughout the Lishui region. Also, over 380,000 households have received the IC Card, which is incorporated with the Rural Credit Database system for credit rating information sharing. As a result, the lead time for forest rights mortgage approval has reduced from 1 week to only half a day, while the average household credit level has increased to 22,500 RMB (Chen & Yao, 2009).

Although the ICT-based financial service platforms have achieved some success in helping rural communities and small businesses in Lishui, there are also issues associated with the current measures. At the moment, the use of POS systems for personal banking in villages still has some limitations. Firstly, store owners differ in their knowledge and ability in helping to deliver banking services. Some owners lack accounting skills to properly maintain transaction records. Secondly, villagers can only perform cash withdrawal and deposit transactions by using the cash cards at the POS machines when both are issued by the same bank, while cards and POS machines from different banks are incompatible. Such an issue is largely due to the protective attitude of banks towards their competing business interests in attracting rural customers and lack of information exchange among proprietary account databases.

While databases for banking information, credit rating and loan records, forestry resources and assets, as well as forest rights registry are separately owned and maintained by different institutions, the local governments in Lishui have played a pivotal role in committing significant amounts of funding and administrative support to the development and implementation of e-platforms for rural financial services across the Lishui region. It is understood that such projects could not have been undertaken successfully by individual commercial banks or financial institutions, as they are either short of resources or lack commercial incentives to establish a region-wide information platform that allows shared access. However, such a development model for digitalisation of rural financial services may be difficult for other under-developed rural regions in China to adopt, due to a very limited government budget and administrative resources. Different types of ICT-based service innovation will be required to meet local needs and capabilities.

7.4 Discussion

As discussed in the case study, the examples of rural financial services in Lishui demonstrate some interesting and unique features in applying ICTs to facilitate contextualised solutions for improved service access and service delivery to rural communities. While a main challenge for rural ICTs and ICT-based rural services is the lack of a structured approach to direct the development of systems solutions, this section seeks to analyse further the Lishui case and to identify some key characteristics that can contribute to the practice of, and theoretical models for, rural ICTs.

7.4.1 ICTs and Service Integration

Essentially, the rural financial service innovation in Lishui is ICT-enabled, service-focussed, and result-oriented. As reflected in the case study, the rural financial service in Lishui includes three main service scenarios supported by ICT appliances and resources, i.e., the POS system for personal banking and the information management systems for forest rights mortgage and microloans. Although using ICT means for rural financial service delivery and access has been widely practised in many developing countries (e.g., mobile banking in Kenya), in most cases only a single ICT platform, such as mobile phone or the internet, has been applied for service delivery; this often falls short due to having limited capability to ensure the quality and reliability of service access for diverse rural users. In the Lishui case, however, multiple ICT platforms are deployed to facilitate manifold solution scenarios for different aspects of the rural financial service. In the personal banking service scenario, the POS machines and peripherals provided by the banks to local convenience stores for money transfer and account management present as low-cost ICT devices. Not only do these enable convenient access to basic banking services for the villagers, they also help the local store owners with improved cash flows. Meanwhile, the databases and web portal are implemented as platforms for conveying financial and capital management information services, which are essential for both local forest farmers and financial institutes in managing loans and risks. A unique feature of this case is that ICT devices and resources are identified and carefully tailored to the service needs and socio-economic context of the rural community, instead of linking possible service scenarios to an extant ICT platform already in place. As a result, the demand for the inputs of local resources and skill/knowledge possessed by the village customers to access the banking and mortgage loan services is minimal in comparison with other similar ICT-based rural banking and financial service cases. For example, using mobile phone and text messaging for personal banking in rural regions is often constrained by limited mobile network coverage and low literacy level of local villagers. It is particularly important for rural communities, as well as for the rural service facilitators, to have the technical means compatible to their needs and capabilities.

Despite some limitations in their implementation, the ICT platforms applied in the Lishui case are well connected to facilitate amalgamation of different services for delivering holistic solutions. The interaction between the databases for household credit and for forest property information supports the integration of both microloan and forest rights mortgage approval and management. Meanwhile, the information platforms for microloan and forest rights mortgage services are also linked with the web platform for forest rights exchange and asset valuation and insurance, so that these services—offered by different providers—can be effectively and efficiently accessed together by the community. Furthermore, these ICT platforms not only support different rural financial service scenarios in delivery, but also help to incorporate other services for local business and entrepreneurship development. For example, the POS system is introduced at village convenience stores to facilitate personal banking service to the local farmers. However, the system also enables extra services for the store owners who can utilise this platform to improve their cash flows, their business and financial management knowledge and skills, as well as their clientele and sales of other goods. Unlike the similar results observed from the Indiagriline (Gollakota, 2008), such co-benefits in the Lishui case are, in fact, attributed to ICT applications that are purposely designed and implemented to integrate multiple services.

7.4.2 *ICTs and Synergy-Based Collaboration*

In the Lishui case, the POS system-based personal banking services and the databases for forestry property rights and forestry asset information were developed with the collaborative effort from local governments and relevant public and private financial institutions. As discussed previously, collaboration between public and private organisations is commonly an essential element in many ICT applications for rural development. Mostly, it is the private organisations that play the leading role in the public-private partnership to initiate projects and to drive the collaboration in implementing ICT platforms and service innovations, as reflected in the examples Indiagriline, the M-PESA project, and Grameen Village Phones. However, the multi-stakeholder collaboration for the ICT-based rural financial service innovation in Lishui demonstrates the characteristics of a different paradigm that effectively leads to a solution-oriented partnership based strongly on synergism and synergistic relationships among different players.

Synergy is often used to describe special effects generated from cooperative relationships and behaviours of several parties or individuals, which normally are different from or greater than what can be achieved by summing the contributions of individuals acting alone (Corning, 1998). According to Buss (1987), synergistic interactions are capable of producing positive or mutually beneficial functional results. In general, the synergism concept aims at the functional effects from cooperative relationships among the system elements, rather than the relationships themselves, as the focus for the development of a system solution. It is often argued that collaboration observed in a complex system is an outcome rather than a cause of synergies. In a synergistic system, interactions among entities (parts or wholes) and individual interests are managed for the common interest and collective benefits of the group (Leigh, 1991).

In the Lishui case, despite their different views and individual interests, the local governments, banks, forestry finance institutions, and community share a common objective to put in place effective and efficient technological platforms and service solutions to provide farmers and small cooperatives with improved access to basic financial services to enable better living and local business development. While these participants all fall short of resources and capabilities to deliver solutions on their own, such consensus with a common goal (as stated earlier) forms the basis for positive synergism among the stakeholders to take collaborative actions in service co-development and co-delivery, which has notable power for innovation. The ICT-based service platforms and service scenarios are developed as the result of synergies among the stakeholders in their capabilities and resources. For example, the POS system for personal banking and money transfer is built upon the synergistic linkages among the administrative capabilities and financial resources of the public institutions (i.e., local governments and the PBC), the business establishment and entrepreneurial competencies of the village convenience store owners, and the service capabilities and technical resources invested by the commercial banks. Similarly, the information systems for household credit and forestry rights mortgage are developed by the collaborative actions and partnership of the Lishui government, the RCC, forestry asset management agencies, and forestry finance agencies. This also developed as the synergistic functional effect of the complementary capabilities and resources among the partner entities in accessing and managing the mission-critical data and information.

Therefore, the ICT solutions deployed in the Lishui case are essentially results of synergies among heterogeneous organisations in their partnership and collaboration based on common goals, capability matching, and resource sharing. Meanwhile, the ICTs also act as a platform for rural financial service collaboration of the stakeholders, and are conducive in realising positive synergism among the participating organisations, their technical and business competencies, as well as the different service scenarios for solution co-development and integration.

7.4.3 ICTs and Product-Service Systems

In many circumstances rural ICT solutions are actually not about, or limited to, the technical means of ICTs. Rather, they are more centred on ICT-enabled services. ICT devices and networks effectively act as media for services that provide a platform to facilitate service access, service integration, and service delivery for rural community development. This is clearly reflected in the Lishui case, as well as in the cases of M-PESA mobile banking service and Indiagriline information services. In this sense, rural ICTs and ICT-enabled services are inherently associated with collaborative endeavour of multiple heterogeneous stakeholders for solution-oriented partnerships, which are built on the synergies among their interests, their roles, as well as their resources and capabilities for value co-creation. While ICT solutions are outcomes of such synergies, ICT platforms also enable collaborative service innovations and development/expansion of the partnership networks. Thus, fundamentally, rural ICTs represent the characteristics of service-based social-technological solutions that are enabled and conveyed through technical ICT systems or products (hardware and/or software). This is essentially in line with the philosophy and paradigm of Product-Service Systems (PSS).

In general, PSS can be defined as an innovation strategy that shifts the business focus from offering physical products alone to offering a system of products and services jointly capable of fulfilling specific demands (Manzini & Vezzoli, 2002). From the traditional systems function perspective, physical or tangible entities (in other words, 'products') are responsible for carrying out the functions of PSS while nonphysical or intangible entities (also known as 'services') ensure the smooth delivery of the functions (Maussang, Zwolinski, & Brissaud, 2009), with products being seen as 'service carriers'. Although PSS applications and their resulting benefits are viewed mainly from business perspectives, in recent years some research efforts have also been made to have the PSS concept extended to the context of regional community development for improving productivity and fostering socio-economic transformation, which leads to a new PSS paradigm. For example, the work of Devisscher and Mont (2008) demonstrates the potential for PSS and system optimisation approaches to improve the integrated economic, environmental and social sustainability for rural farming communities in a developing nation.

The notion of an effective PSS for community transformation can be captured by interlinked building blocks of products, services, networks of actors and partners, and enabling mechanisms for driving and managing value co-creation through the analytical framework for PSS solutions presented in Fig. 6.1 in the previous chapter. Such a framework can assist the identification and development of contextualised PSS scenarios and the opportunities for their integration. By linking the PSS framework with the Lishui case, the characteristics of the case contexts, the participants and partnerships, ICT-based rural financial service innovation scenarios, and their interconnections can be interpreted and depicted in Fig. 7.2.

- Actors: As discussed in the previous section, the main actors for the rural financial service innovation in Lishui are banks and financial institutions, public agencies, local governments, as well as local rural communities (e.g., farmers and grocery store operators). Among them, the PBC and the local Lishui governments play the key role as the *facilitators* who drive and coordinate the service innovation projects and partnerships among different actors with divergent interests.
- Products: According to the framework, as carriers of services products can be either physical or intangible goods that are tradable. They can also include capital goods, such as machinery, equipment, or facilities, for production or business operations. In the Lishui case, it is clear that the local forest products and forest

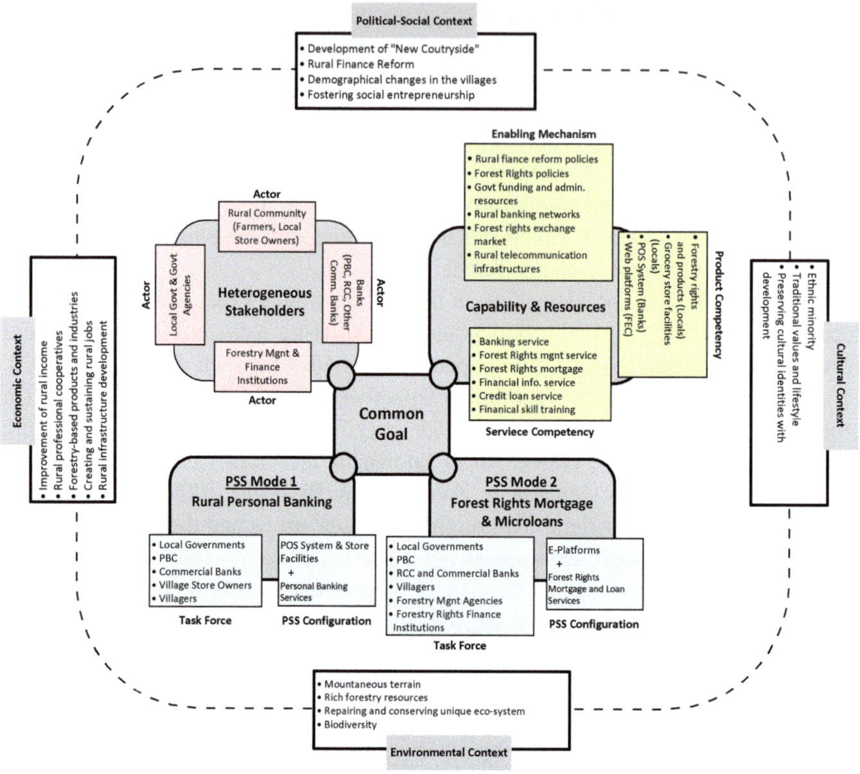

Fig. 7.2 ICT-based rural finance service innovation in Lishui as a synergy-based PSS development

property rights are key tradable products (or commercial goods) that underpin the forest rights mortgage and microloan financial services. They are the key competencies of the local farmers. Meanwhile, the local grocery store facilities, the POS devices, and the ICT platforms (e.g., the online web portals for forest rights trading) are considered as critical capital products deployed for generating and delivering the expected banking and mortgage loan services to the local farmers and small businesses.

- Services: As the main impetus for realising the rural economic development in Lishui, the suite of services to improve the access of local communities to financial capital include personal banking, credit assessment and management, forest asset evaluation and forest rights registry management, credit loan and mortgage provision, financial and forest rights exchange information, as well as financial skill training for local shop owners. These services are derived from the service competencies of the actors (e.g., banks, financial institutions, public agencies, and local businesses). The desired outcomes are achieved through appropriate combinations of these service items and integrating with the right 'products'.

- Enabling mechanisms: The rural financial service innovation in Lishui is fundamentally driven by the central government's rural finance reform policies, including the policies on forest rights reform and forest rights trading, and regulated by the political establishment of the local governments. These are two critical mechanisms for structuring governance of the development. Another key enabling mechanism for the ICT-based service innovation is the rural ICT infrastructure, including telecommunication and broadband networks, which provide indispensible technical means for information sharing and service access. In addition, funding and administration resources from the public and private sectors enable the setup and implementation of the databases and information management systems for the financial service delivery. Also part of the main enabling mechanisms are the rural operation networks of the participating commercial banks, mortgage loan providers, and forest rights agencies in the Lishui region for conveying financial services, as well as the exchange market for trading forestry resources and property rights to generate capital for forestry farmers and businesses.
- PSS scenarios: A PSS scenario is a highly contextualised solution that fits with and responds to characteristics of external environments, which present both as constraints and as opportunities/resources for the service innovation. The natural, economic, socio-political, and cultural conditions of the Lishui case were examined earlier and summarised as shown in Fig. 7.2. Through capability matching among the heterogeneous actors and exploring the synergies between the product and service competencies and the community needs, POS system-based personal banking and e-platforms for credit loans and forest rights mortgages are developed as two ICT-based PSS scenarios for improving access to rural finance. Each of the PSS scenarios consists of a cluster of pertinent actors as a focused taskforce and an integration of suitable products, services and enabling mechanisms to form an innovative product-service configuration to cater for particular service needs. However, through a holistic planning and systematic design process, it is possible to capture the opportunities to connect different scenarios together for shared, and more efficient, use of infrastructures and resources.

While the PSS paradigm for community development is intended for broader development of product-service solutions, as demonstrated by the discussion above, it is certainly applicable for ICT-enabled service innovations and can address the key characteristics of rural ICTs that are identified from the Lishui Case. The PSS model presented can be adapted to provide a systematic process for the development and implementation of rural ICT solutions, in particular ICT-based rural service innovations, which are currently derived more from ad hoc or experience/example-based approaches. Although PSS is a relatively established concept, its application for community transformation is not yet fully examined and utilised. It is evident that further research, supported by empirical case studies, is necessary to further investigate the mechanisms for applying PSS paradigm and design methodologies to underpin service-oriented rural ICT development and implementation.

7.5 Summary

The ability of rural communities in developing nations to improve their livelihood and productivity is largely limited by the lack of access to information services and information assets essential to their day-to-day needs. It has been increasingly recognised that ICTs play a vital role in capacity building and improving quality of socio-economic wellbeing for rural empowerment. While ICT solutions for rural development are often technologically focused on the universal access to devices and infrastructures, the review of extant rural ICT examples show that ICTs are means rather than ends in such contexts, and it is more useful for farmers and small rural businesses to gain universal access to enabling services through low-cost ICT platforms.

Despite much research and practical effort having been devoted to ICT-enabled services for rural development, it is found that current approaches largely rely on imitating extant examples or learning from best practices. A major limitation in rural ICT applications is the lack of a clear and systematic model to capture the characteristics of rural ICTs and to guide the development of collaborative partnership and systems solutions for ICT-based service innovation. To address such issues, a detailed analysis on an ICT-enabled rural financial service case in Lishui, China has been presented to identify some key features and paradigms for rural ICT service innovation. The Lishui financial service innovation enables improved local accesses to personal banking, forest right mortgages, and microloans with the support of ICT networks and ICT service solutions. POS systems at village convenience stores and information systems for forestry asset and forest right management were developed as innovative ICT platforms to facilitate delivery of the basic financial services to the rural forest farmers and small forest cooperatives in Lishui.

As reflected in the Lishui case, ICT-enabled rural services demonstrate a strong element of multi-stakeholder collaboration and collaborative innovation, which are facilitated by public institutions (e.g., local governments in Lishui and the central bank) and involve local community participation as well as heterogeneous business partners (e.g., various commercial banks and forestry capital service providers). Rural ICT solutions are developed as the result of a synergy-based partnership between different organisations with shared goals, interests, capabilities and resources. In addition, ICTs are not only outcomes of collaboration, but also the platform to facilitate collaborative service innovation and to help integration of different services for holistic solutions. Essentially, ICT devices, applications, and infrastructures are embodiments or carriers of rural services. Such ICT-enabled service innovation characteristics are essentially in line with the PSS concept and paradigm. Based on the case analysis, it is argued that a synergy-based PSS model for community transformation can be adopted to provide a systematic process to guide the rural ICTs and ICT services development. This proposition and the correlation between PSS and rural ICTs need to be further examined and developed in future research work, with more empirical studies on rural ICTs and ICT-enabled service innovation cases.

References

Anand, A., Pejovic, V., Johnson, D., & Belding, E. M. (2012). VillageCell: Cost effective cellular connectivity in rural areas. In *Proceedings of the 5th international conference on information and communication technologies and development* (pp. 180–189). ACM-SIGCHI

Bakhshizadeh, H., Hosseinpour, M., & Pahlevanzadeh, F. (2011). Rural ICT interactive planning in Ardabil province: Sardabeh case study. *Procedia Computer Science, 3*, 254–259.

Barrett, M., & Slavova, M. (2011). *Making ICT infrastructure, appliances, and services more accessible and affordable in rural areas*. ICT in agriculture: Connecting smallholders to knowledge, networks, and institutions – Module 2 (pp. 15–48). The International Bank for Reconstruction and Development, The World Bank.

Black, R., & Atkinson, J. (2007). Addressing the digital divide in rural Australia. In *13th Australaasian world wide web conference, AusWeb'07* (pp. 45–62). Australia: Southern Cross University Printery.

Brouwer, R., & Brito, L. (2012). Cellular phones in Mozambique: Who has them and who doesn't? *Technological Forecasting & Social Change, 79*, 231–243.

Buss, L. W. (1987). *The evolution of individuality*. Princeton, NJ: Princeton University Press.

Caspary, G., & O'Connor, D. (2003). *Providing low-cost information technology access to rural communities in developing countries: What works? What pays?* (Working Paper No. 229) (p. 48) OECD Development Centre.

Chapman, R., & Slaymaker, T. (2002). *ICTs and rural development: Review of the literature, current interventions and opportunities for action* (Working Paper 192 for Overseas Development Institute).

Chen, D., & Yao, Y. (2009). Innovation for forest right mortgage loan mode in Qingyuan: A "forest right IC card" brings convenience to 40,000 forest farmers. *Zhejiang Online*. Retrieved March 10, 2013, from http://zjnews.zjol.com.cn/05zjnews/system/2009/04/21/015445015.shtml (in Chinese).

Chen, J. (2012). Investigation on continuous improvement of "Forest Right IC Card" system in Lishui. *China Finance Online*. Retrieved April 12, 2013, from http://yhcs.bank.cnfol.com/110808/138,1398,10430168,00.shtml (in Chinese).

Corning, P. A. (1998). The synergism hypothesis: On the concept of synergy and its role in evolution of complex system. *Journal of Social and Evolutionary Systems, 21*(2). Retrieved February 8, 2011, from http://www.complexsystems.org/publications/synhypo.html

Devisscher, T., & Mont, O. (2008). An analysis of a product-service system in Bolivia: Coffee in Yungas. *International Journal of Innovation and Sustainable Development, 3*(3/4), 262–284.

Dymond, A., Oestmann, S., Whiting, K., Smithers, C., Milne, C., & Milne., R. (2010) Module 4: Universal Access and Service. In *ICT regulation toolkit*. InfoDev, Retrieved on May 10, 2013, from http://www.ictregulationtoolkit.org/en/Sections.html

Enejo, I. F. (2009). Strategies for low cost rural telephony. *The Pacific Journal of Science and Technology, 10*(1), 256–261.

Essegbey, G. O., & Frempong, G. K. (2011). Creating space for innovation – The case of mobile telephony in MSEs in Ghana. *Technovation, 31*, 679–688.

Futch, M., & McIntosh, C. (2009). Tracking the introduction of village phone product in Rwanda. *Information Technologies and International Development, 5*(3), 54–81.

Gollakota, K. (2008). ICT use by businesses in rural India: The case of EID Parry's Indiagriline. *International Journal of Information Management, 28*, 336–341.

Gu, L., & Wang, Y. J. (2012). Why farmers in Lishui can withdraw cash from POS machines. *Chinese Economics Times*. Retrieved March 12, 2013, from http://zjnews.zjol.com.cn/05zjnews/system/2009/04/21/015445015.shtml (in Chinese)

James, J. (2011). Sharing mobile phones in developing countries: Implications for the digital divide. *Technological Forecasting & Social Change, 78*, 729–735.

Karugu, W. N., & Mwendwa, T. (2007). Vodafone and Safaricom Kenya: Extending the range and reliability of financial services to the poor in rural Kenya. Retrieved February 12, 2013, from *Growing Inclusive Markets*, http://growinginclusivemarkets.org/media/cases/Kenya_MPESA_2008.pdf

Kiiza, B., & Pederson, G. (2012). ICT-based market information and adoption of agricultural seed technologies: Insights from Uganda. *Telecommunications Policy, 36*, 253–259.

Leigh, E. (1991). Genes, bees and ecosystems: The evolution of a common interest among individuals. *Tree, 6*(8), 257–262.

Mackie, N. (2010). Facebook for farmers: Technology empowers China's rural workers. *BBC News: Business*. Retrieved October 23, 2012, from http://www.bbc.co.uk/news/business-12010549?print=true

Manzini, E., & Vezzoli, C. (2002). *Product-service systems and sustainability: Opportunities for sustainable solutions*. Paris: UN Environment Programme (UNEP), Division of Technology, Industry and Economics, Production and Consumption Branch.

Maussang, N., Zwolinski, P., & Brissaud, D. (2009). Product-service system design methodology: From the PSS – Architecture design to the products specifications. *Journal of Engineering Design, 20*(4), 349.

Mayeku, B., Kilwake, J., & Bertarelli, F. (2010). Use of low-cost wireless communication technology for education in rural Kenya. In *Proceedings of the 4th ACM workshop on networked systems for developing regions*, Article 13, doi:10.1145/1836001.1836014

NECTC. (National Electronics and Computer Technology Center). (2003). *IT for poverty reduction: Sample cases from Thailand*. Retrieved on May 10, 2013, from http://www.nectec.or.th/pub/book/itforpovertyreduction.pdf

Rao, N. H. (2007). A framework for implementing information and communication technologies in agricultural development in India. *Technological Forecasting & Social Change, 74*, 491–518.

Ratsifandrihamanana, L. H. (2012). Cooperatives and the role of information and communication technologies (ICTs) – Panel Discussion Note, The Division for Social Policy and Development (DSPD) of the United Nations Department of Economic and Social Affairs (UN-DESA). Retrieved January 12, 2013, from http://social.un.org/coopsyear/documents/FAO-speech-cooperatives-ict.pdf

Rugonzibwa, P. (2012). *Tanzania: 48bn/- set aside for rural telecoms*. Retrieved on March 12, 2013, from http://kenyadailyeye.jibostudios.com/2012/10/tanzania-48bn-set-aside-for-rural-telecoms/

Shi, Y., Lin, Y., & Zhou, Y. (2011, December 16). Forestry credit assurance: Financial services for the forestry industry to facilitate 'green growth'. *Zhejiang Daily, 8*.

Tengtrakul, P., & Peha, J. M. (2011). Access to and penetration of ICT in rural Thailand. *Telecommunications Policy, 35*, 141–155.

Tobgyl, T. C. (n.d.). *Case study: Bhutan – Wireless IP based Rural Access Pilot Project*. Retrieved February 23, 2013, from http://www.itu.int/osg/spu/ni/futuremobile/technology/bhutancase.pdf

Tsao, H., Jacob, S., Parikh, S., Ghosh, A., Pal, R., Ranalkar, M. et al., (2010). Streamlining grain supply chains of India: Cloud computing and distributed hubbing for wholesale-retail logistics. In *Proceedings of 2010 IEEE international conference on service operations and logistics and informatics* (SOLI) (pp. 252–257).

Wang, Y. J., & Gu, L. (2012). What can be learnt from the "Lishui Model". *Chinese Economics Times*. Retrieved April 14, 2013, from http://news.xinhuanet.com/fortune/2012-07/23/c_123453198.htm (in Chinese).

Zhang, J. H. (2012). Come forth the "Lishui Model". *Caijing Magazine*. Retrieved on April 14, 2013, from http://www.chinareviewnews.com/crn-webapp/mag/docDetail.jsp?coluid=0&docid=102302208 (in Chinese).

Chapter 8
Towards an Integrated Model

8.1 Introduction

Drawing together the themes and proposals from the preceding chapters, we now put forward an integrated model for community transformation. With stronger communities at its core, the model recognises the key role of social enterprises in job creation and community transformation, supported by various actors working together, combining their capabilities and sharing assets and resources in a synergistic, effective and efficient manner. A methodical planning process is then described, leading to a systematic planning framework for community transformation that also reflects the key elements of the model. The chapter also sets out a direction for further research, including on the ground application and testing of the integrated model, process and framework, draws on learnings from Australia and Malaysia, and discusses implications of the approach to other communities in the region and beyond.

8.2 Summary of Findings

With the goal of enabling stronger rural communities, as outlined in Chap. 1, this book has examined how social entrepreneurs could form community enterprises, generating employment, improved incomes and livelihoods, when supported by infrastructure systems, financial instruments and investment, and partnerships with the corporate sector.

In Chap. 2, we identified that individual social entrepreneurs performed a central role in establishing and operating community enterprises. Based on a study of six cases, we concluded that multiple key characteristics are necessary to enable leaders to develop rural cooperatives that can unite rural communities in achieving common prosperity.

Y. Zhu et al., *Transforming Rural Communities in China and Beyond*,
DOI 10.1007/978-3-319-11319-7_8

Chapter 3 introduced one of the key components of improving rural prosperity and reducing greenhouse emissions, namely, a market-based financing mechanism and a voluntary trading platform for Chinese forest carbon sinks. The pilot program involving the Huadong Forestry Exchange was described, with a focus on its financing mechanism, the China Green Carbon Fund. In order to develop a voluntary forest carbon market in China, with sustainable forest carbon sink projects, a highly innovative and flexible and market-based financing mechanism will be required. Possible implications for other provinces and local areas were outlined, and the implications for other developing countries that are introducing trading platforms for forest carbon sinks.

Chapter 4 explained the potential for trading in forest land rights and carbon sinks to improve livelihoods for rural communities and diversify income sources. The role of village cooperatives in negotiating with afforestation companies for carbon credit schemes was highlighted, with improved community infrastructure being negotiated as part of the agreement. Individually, farmers struggled to handle the complexities and were at a disadvantaged position. Thus, farmers may become shareholders through forestry rights investment, with afforestation companies being responsible for investment on afforestation, providing technologies, management and bearing natural disaster and investment risks.

The discussion of alternative energy development in Chap. 5 revealed that effective energy access involved not only technologies but also ‘soft systems’ and energy services. An interconnected ‘whole system approach’ could offer affordable energy, pollution prevention and climate change mitigation. Furthermore, in the case of off-grid electrification programs and especially small hydropower projects, public-private partnerships could augment local cooperatives to support local economies while mobilising capital from the private sector.

Chapter 6 illustrated how synergistic, integrated and inclusive planning of services and infrastructure could lead to ‘win-win’ solutions that result in improved community outcomes with less resource use and less cost. The various actors can share their capabilities, resources and infrastructure in pursuit of various transformation themes or scenarios, enabling interrelated products and services, such as tea and organic food services, ecotourism, and the like, to be delivered more efficiently and effectively.

Chapter 7 illustrated this approach in more depth, by means of a case study of ICT-enabled rural financial services in Lishui, Zhejiang Province. It was shown how synergies among heterogeneous organisations, with common goals, capability matching and resource sharing, could improve access of remote rural communities to personal banking, microfinance and forest rights financial services. ICTs constituted a platform for collaborative and integrated, holistic service innovation. Thus, a synergy-based model for community transformation was applied, which could lead to development of a systematic process.

The book also sought to develop a model for building stronger communities that become self-sufficient over time. This model should recognise key elements such as: self-determination; creation of social enterprises; fostering of social entrepreneurs and leaders; optimisation of local skills, knowledge, assets and natural resources;

partnerships with external supporters, investors and advisors; and governance mechanisms that actively involve community members and actors.

The following sections explain the development of such a model and associated processes.

8.3 Toward an Integrated Model for 'Stronger Community'

As explained in the first chapter and depicted by Fig. 1.1, the primary focus of the group mission project is on how social entrepreneurship and community enterprises, supported by appropriate and integrated infrastructure systems, enabling technologies, and financial models, can help develop stronger rural communities. The concept of 'building a stronger community' is based upon empowering and assisting the community to achieve resilience and self-dependence. Based on the investigations and findings reported in the previous chapters, some important attributes contributing to a stronger and more resilient community include:

- strong mentality of self-determination by the community, controlling its own destiny and setting its own goals;
- strong champions and leaders that draw together individual social entrepreneurs and enterprises under cooperative mechanisms, with a common village brand and identity;
- recognition of village context, such as demographical, cultural, natural environments; skills, ideas and capacities; and associated challenges and opportunities;
- shared utilisation and integration of village capabilities and resources, including facilities and infrastructures, unique natural assets and ecosystems, and ideas and human talents;
- partnership of the community with external public and private institutions, corporations, scholars and NGOs, to gain capital and investment, capability building, education and training, and technical, business and financial advice and support; and
- government policy support, such as by fostering cultural identities of ethnic minorities (e.g., the *Bulang* and *She* people), and promoting conservation of traditional architecture and crafts.

These attributes also represent the key elements that need to be properly incorporated and addressed in the community development process to deliver multiple win-win outcomes, including individual benefits, community prosperity, profitable enterprises, and conservation of the environment and local culture. Therefore, an integrated development model supported by a strategic planning approach is needed in order to combine these elements systematically with other aspects of community empowerment.

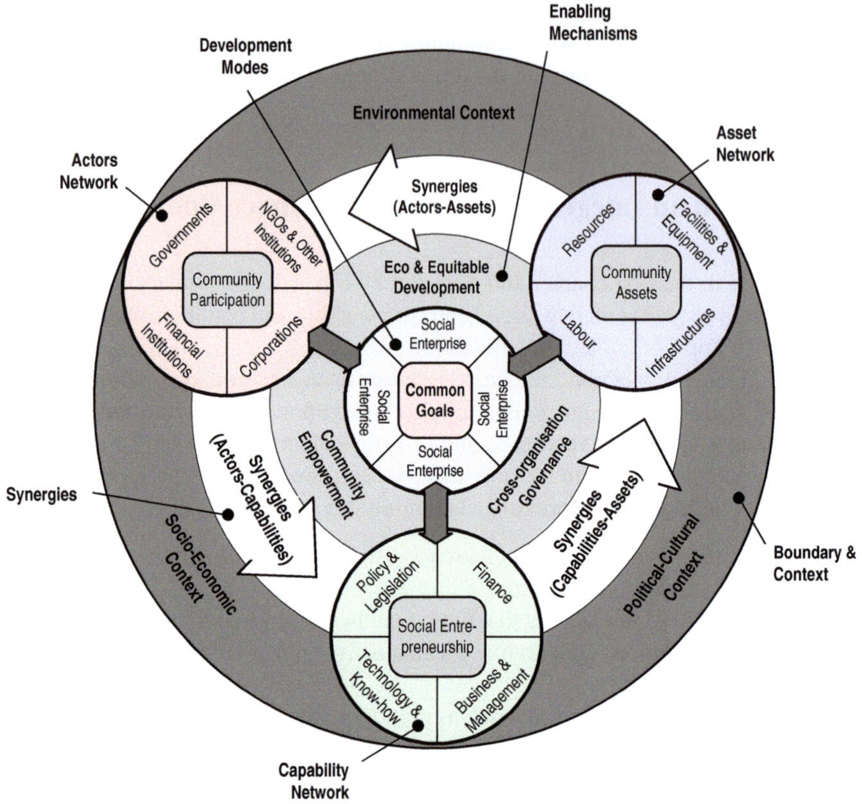

Fig. 8.1 Integrated model for development of stronger community

8.3.1 The Construct of an Integrated Model

Essentially, the development of a rural community has to recognise the needs and ability of the community and to foster social enterprises and entrepreneurship through the partnership with external organisations and corporate sector. It is critical to have a common goal and contextualised socio-technical-business solutions supported by connecting interests, resources and infrastructures, and capabilities among heterogeneous stakeholders. Based on such a notion and the paradigms for synergy-based community transformation discussed in Chaps. 6 and 7, the following conceptual model (Fig. 8.1) can be applied to generate systems solutions for desired transformation outcomes.

The construct of the model takes the structure of 'planets and orbits' to capture and integrate the key elements as well as their interconnections. As indicated in the figure, the model consists of eight major components, i.e., Boundary and Context, Synergies, Enabling Mechanisms, Networks of Actors, Assets, and Capabilities, Development Scenarios, and Common Goals.

- *Boundary and Context*: Solutions for transforming a rural community need to fit with the characteristics, strengths, and needs of the community. Identifying and assessing the environmental, economic, cultural-political, and socioeconomic conditions of the case community are critical for any such development project to investigate the issues, constraints, and opportunities for implementing contextualised solutions. Establishing proper and adequate understanding about the community context is also necessary for setting clear and specific system boundary and scope for the projects, within which relevant elements and their interactions can be further defined, examined, and manipulated.
- *Common Goals*: Central to this mini-universe, the common goals represent the ultimate focus as well as outcomes to achieve a stronger community. These goals can include a set of distinctive, yet mutually reinforcing, objectives of transformation. The goals are defined based on aligning and matching the views and the interests of the community and other stakeholders for common grounds, while recognising and reconciling their differences, with respect to the given context. Preferably, identifying and defining such common transformation goals should lead to a goal statement. For example, in the case study on the ICT-based rural financial service innovation in Lishui (in Chap. 7), "developing viable and easy-to-access platforms for low-cost banking and capital services to rural communities and small cooperatives" is stated as a common goal that is derived by drawing together the needs and interests of community, local governments, banks, and financial service providers.
- *Development Modes*: Around the common goals for stronger community, there is a cluster of 'social enterprises', each representing a special development mode, or scenario, for establishing cooperative or social business entities toward human and environmental well-being of the community in achieving the transformation objectives. Multiple development modes may be constructed to address value propositions and objectives shared by various groups of stakeholders and to specify pathways in reaching the stronger community goals. With conceptualisation of the modes, co-design is carried out in detail to substantiate the social enterprise development. As discussed in Chap. 6, there can be different types and forms of social enterprises in generating various goods and services, such as forestry products, agricultural and horticultural products, tea production, farmhouse tourism, hydro-electricity generation, logistics and supply chain services, etc. Each social enterprise scenario can be considered individually and then holistically within the overall goal of achieving community transformation.
- *Actor Network*: The network of actors is a consortium of heterogeneous stakeholders, such as people in the community, corporations, government agencies, financial institutions, NGOs, and other entities (e.g., research and education institutions, professional associations, healthcare service providers, etc.), that have similar vision and intention for value co-creation to engage with the identified challenges and to achieve the desired transformation. Central to this network is the community and its direct participation in decision-making and implementation. Given the interdependency of the challenges for development that need to be addressed, problem analysis and cross-fertilisation of ideas are carried out

across the disciplinary sectors and professional expertise of the actors. Different views and interests are examined, discussed, and aligned to find overlaps and to link with the community needs. Such a dynamic process can contribute to defining and refining the stronger community goals. Also, special task forces (or sub-networks) with dedicated champions and facilitators can be formed for respective social enterprise projects. On the other hand, the development and implementation of a social enterprise project may also require adjustment of constituents and/or roles within the actor network, leading to its expansion, or sometimes contraction, with involvement of new participants from external supporters, investors, and advisors.

- *Capability Network*: The development of social enterprises not only depends on the involvement of different actors, but also relies on how they utilise what they have (i.e., assets, ecological resources) and what they can do (i.e., capabilities). Basically, the capability network represents a collection of major competences of the respective actors regarding the kinds of knowledge, information, skills, and services that they can offer for the development of social enterprises. A rural community can form such a network—their entrepreneurial knowledge and skills may be creative, innovative, and productive, enabling them to set up and/or operate their own social enterprises to generate and sustain local income and employment. Developing, using, and improving social entrepreneurship in the community needs to be fostered and facilitated by policies, regulations and legislation, financial mechanisms for investment, support and skill training for business management and marketing, as well as appropriate technology and tacit knowledge, which draw upon the competencies of other participants in the actor network.
- *Asset Network*: In addition to matching the competencies possessed by different actors in the capability network, the asset network is to connect different kinds of resources (e.g., ecological, material, financial, human, etc.), infrastructures (e.g., transport, utility, telecommunication, etc.), facilities and equipment, and various forms of goods and properties possessed by different actors to support the establishment and operations of social enterprises. Core to the community's role in achieving transformation goals is mobilising people in the community and combining their entrepreneurial skills, practical knowledge, natural and human resources, as well as physical assets (e.g., forestry land, tea plantation, agricultural produce, manufactured goods, etc.) and non-physical assets (e.g., financial capital, credit, cultural heritage, etc.). Through the partnership developed in the actor network, the local knowledge and assets of the community can be effectively linked with the resources and skills of local government, local businesses, and civil society entities to facilitate the search for and experimenting with new opportunities/practices for improvement. A social enterprise is effectively a special clustering of a task force of actors (including the community), their capabilities, and their assets. Different social enterprise scenarios and entities may have overlaps in actors involved, assets used, and capabilities applied. Each scenario needs to be considered individually and then holistically within the overall goal of achieving stronger community transformation.

- *Enabling Mechanisms*: In addition to the actor, capability, and asset networks, establishing and running social enterprises is also based upon and enabled by a series of operating and managerial strategies and models, also known as enabling mechanisms. In essence, enabling mechanisms define 'who does what', 'when', and 'how' in the co-creation and co-operation of social enterprise entities through cross-organisation governance structure to enforce strategies for community empowerment, with models of ecological and equitable development. For a social enterprise, a cross-organisation governance structure in place can manage the dynamic processes of partnership (i.e., forming, evolving, deepening, diversifying, terminating, etc.), coordinate information exchange and resource sharing, and regulate collaboration (in deploying and integrating capabilities and assets) behaviour among the actors in the task force. Meanwhile, *community empowerment* strategies derived by connecting actors and their capabilities and skills are implemented to help enhance the effectiveness of community participation and improvement of social entrepreneurship. Such strategic measures include skill development, supportive legislative and political processes, effective financial services, and efficient technological platforms that foster local champions, leaders and entrepreneurs, as well as nurture local skills, ideas and knowledge to make the community self-sufficient and self-sustaining over time. In addition, *ecological (eco) and equitable development* represents particular policy and business models for setting up and operating social enterprises that are fair and respectful to the interests of all stakeholders, particularly to the social, cultural, economic, and environmental contexts and needs of the local community. Through an appropriate eco and equitable development model, community assets and natural resources are incorporated and shared in the social enterprise development, as well as being respected, conserved and strengthened for sustainable use and development.
- *Synergies*: As discussed in Chap. 7, the concept of synergy focusses on interactions among entities and assumes that synergistic interactions are capable of producing positive or mutually beneficial functional outcomes where individual interests/capabilities are managed for the collective benefits/capabilities of the group. While the enabling mechanism elements are interconnected, mutually enhancing, and assist the development of social enterprise scenarios and entities, the governance structure, empowerment strategies, eco and equitable development models are inherently built upon and scaffolded by synergies within and among the networks of actors, assets, and capabilities to support the formation and operation of social enterprises. Fundamentally, the construct of this integrated model represents a synergistic system that explores and facilitates opportunities for collaborative partnership and solution co-development among different stakeholders through interest alignment (synergies in the actor network), capability matching (synergies in the capability network), and connection of various forms of assets and resources (synergies in the asset network).

8.3.2 *Systematic Planning Process*

The model presented in Fig. 8.1 provides an integrated and holistic view of how different stakeholder knowledge, skills, resource elements and their interactions can be captured and contribute to the development of social enterprises and to addressing the key attributes of stronger community. However, this integrated model itself does not present an executable road map to attain the desired outcomes and to achieve the community transformation goals. Therefore, a methodical planning process needs to be applied to enable the model implementation.

The UNESCAP (2011) has outlined a strategic method for synergistic infrastructure planning and development, which involves mobilising different views and resources towards a common vision, goals and objectives, with careful consideration of various win-win solutions. As illustrated in Fig. 8.2, the four-stage, ten-step approach addresses four essential questions:

1. *Where are we now*? – this leads to understanding the context and problems, to identifying stakeholders, legal frameworks, drivers and barriers, to evaluating capacities, and to performing initial assessment on the current situation.
2. *Where do we want to get to*? – this question can be answered by establishing among stakeholders a common vision for the development and defining a set of objectives and performance indicators related to the vision.

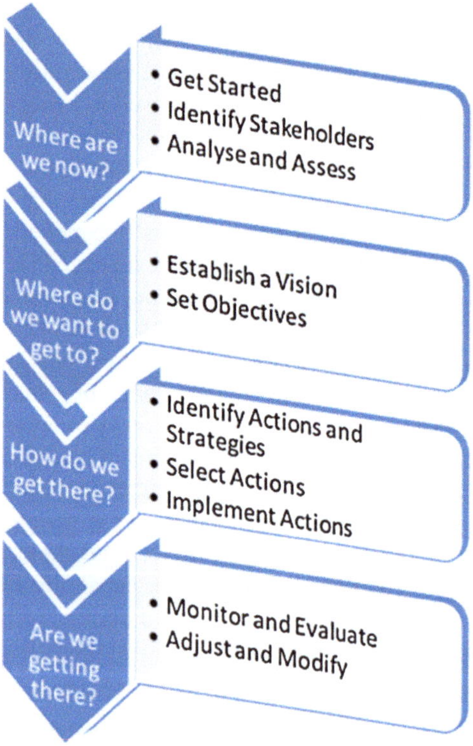

Fig. 8.2 Four-stage, ten-step strategic planning process (Based on UNESCAP, 2011)

3. *How do we get there?* – appropriate strategies and actions will be developed and prioritised, which will also require identifying and addressing inter-organisational coordination and governance gaps for a proper action plan.
4. *Are we getting there?* – this requires monitoring and evaluating the outcomes of the action plan against the set objectives for adjustment or further improvement.

Whilst the UNESCAP planning approach is directed at urban areas, the key concepts can be adopted and connected with the integrated model. Thus, a more structured methodological framework for a systematic planning process can be developed as shown in Fig. 8.3. Similarly, this framework also consists of four stages and 16 steps of planning process to assess the current situation (i.e., Where are we now?), to identify the objectives and desirable results of transformation (i.e., Where do we want to get to?), to devise viable options for development (i.e., How do we get there?), and to evaluate and review the outcomes (i.e., Are we getting there?). As indicated by the framework diagram, some of the planning steps can be involved in different stages.

To assess the current situation of the community of interest, the process starts with gaining a preliminary understanding about the community with regard to its geographicallocation,itsdemographicalprofile,itssocio-economic-environmental-cultural assets and constraints (i.e., haves and have-nots), as well as its key challenges and needs for development. By using such information and knowledge, potential stakeholders of different types and from different sectors are identified and convened

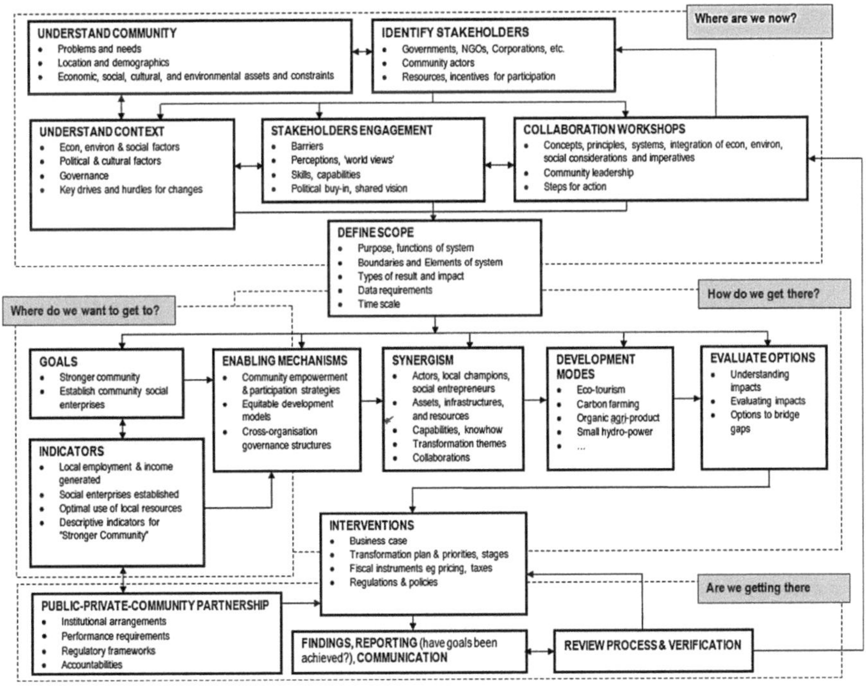

Fig. 8.3 Systematic planning framework for community transformation (Based on Ness, 2009)

based on their willingness to participate and the resources and capabilities that they can bring to address the issues and needs of the community. Together with direct involvement of the people from the community (e.g., local residents, community champions/leaders, social entrepreneurs, etc.), and the internal and external stakeholders (e.g., local governments, public agencies, NGOs, private organisations, businesses, investors, etc.) form a consortium for further communication and engagement to discuss and to develop deeper and broader understanding about how the community's structure and functions are governed and the context for collaboration, by examining critical economic, environmental, social, political, and cultural factors that affect all parties and their roles in the collaborations. Also, it is important for the stakeholders to identify and agree upon the key drivers and barriers to make changes that are desirable to them, in particular to the people of the community.

Structured and facilitated workshops may be organised for the stakeholders to exchange views and ideas, to share information and knowledge, to match interests and capabilities, to converse perceived benefits, barriers, and risks for collaboration, and to identify the forms of resources, mechanisms, and commitments (technical, fiscal, business, as well as political and legal) necessary for dealing with the challenges. Through such an interactive process, new ideas can be created, various concepts can be developed and examined, different considerations and imperatives of individual participating entities as well as their overlaps can be discussed and integrated, community opinions can be expressed and reviewed, and community leadership can be fostered by following some in-built training programs. Subsequently, a series of tasks for partnership building and solution development will be defined and carried out. By conducting and completing these collaborative activities, as an important outcome of this planning stage, the scope for the multi-stakeholder collaboration, the boundaries and the key elements of intended systems solutions, the required data and information to be collated, and, critically, the timeframe for making changes and delivering results are collectively defined and agreed. Different or conflicting opinions and interests are respected and reconciled, which lead to defining the specific targets, measures, and action plans for achieving the desired community development and associated collective benefits.

Similar to the generic planning approach implemented by UNESCAP, the quest to define what kinds of transformation the community should aim to advance and what outcomes are expected requires setting clear goals that reflect the common vision and objectives, which can be resilient over time. Also, the desired outcomes for setting up community-based and community-oriented social enterprises are determined. Meanwhile, by consulting with all stakeholders involved in the process, specific and assessable indicators are decided to define stronger community targets in the particular context and to measure the extent and effectiveness of how transformation goals are met. Key indicators may include how many social entrepreneurs are nurtured, how many social enterprises are set up, how many local jobs are created and sustained, how much local incomes are improved, to what extent local resources and infrastructures are enhanced and optimally used, as well as the degree of community self-sufficiency and level of social capital development. Coupled with setting common goals, a suite of strategies for empowering the community, models for ensuring ecological and equitable development, and governance structures for managing inter-organisation collaborations are also planned as the enabling mechanisms to address and incorporate the defined indicators.

Following the definition of goals and scope, the planning for solution development aims to find answers to how we can achieve the objectives of transformation for stronger community. The strategies, models, and governance structures encompassed in the enabling mechanisms assist in uncovering new opportunities, innovative concepts, and feasible options for development modes that can lead to creation of various social enterprises. Therefore, as advocated by the model, a key part of this solution planning stage is to identify and capture synergistic relationships within, as well as among, the networks of these elements for interest alignment, capability matching, and connection of various forms of assets and resources. From such exercises, a number of potential development modes (such as ecotourism, forest carbon sequestration, organic farming, etc.) can be derived as solution concepts aligned with community empowerment strategies and equitable development models for realising the transformation targets. Various concepts and options for social enterprise development need to be assessed with regard to what and how much impacts they have on the community and the participating stakeholders in meeting the defined transformation indicators (or targets). Therefore, both the gaps of and overlaps between the proposed options of development mode in taskforces, capabilities, infrastructures, and resources can be investigated and recognised so that new and additional options could be developed to provide interconnected and more holistic solutions.

After concept development and evaluation, more structured and concrete action plans for the 'interventions', i.e., building actual business cases for social enterprises, can be generated. For each selected option, the business case may include a detailed timeframe, a specific road map with priorities for step-by-step implementation, as well as fiscal and regulatory arrangements required to support the enterprise development. To ensure the solutions for social enterprise establishment are capable of improving social, economic, environmental, and cultural well-being of the community and addressing the respective bottom-lines of the public and private stakeholders, monitoring and reporting mechanisms need to be set up to assess the performance of the interventions and their impacts against those key indicators of stronger community goals. Meanwhile, in addition to building business cases, it is also critical to forge strong and productive partnerships involving the community, public institutions, and private organisations for co-creation and co-delivery of business and technical solutions. In this sense, the strength and effectiveness of the public-private-community partnership underpinning each intervention also needs to be continuously observed and evaluated with regard to institutional arrangements, regulatory and management frameworks, accountability of partners, as well as other performance requirements governing collaborative behaviour. Through such review and verification, the data and information gathered from the process can help all stakeholders involved in decision-making (including community representatives and community leaders) to clearly understand where they are up to in the development, whether and to what extent the transformation goals have been achieved, and what gaps remain for further changes and improvements in the business cases, in the public-private-community partnerships, in the solution concepts, or even in defining the goals and indicators. Such evaluation and findings can be reported and fed back to help revisit other planning steps and make necessary changes to concept generation, scope definition, solution development, and intervention plan implementation.

8.4 Applying the Model, Process and Framework

Based upon the findings of the various chapters, an integrated model (Fig. 8.1), planning process (Fig. 8.2) and planning framework (Fig. 8.3) have been put forward. Further empirical research is required to test and verify these propositions, by revisiting some of the communities in China or by collaborating with other communities in China or elsewhere.

It will be especially important to engage respectfully with communities and to assist them in taking leadership of their community transformation process. For example, Chap. 6 described various synergistic scenarios for improving the prosperity of the Mangjing, Daxi and Baisha villages; it was emphasised that these need to be explored more deeply with the villagers, while affording them the opportunity to express their own, possibly different, ideas.

Daxi village, Jingning, provides a useful example of a community that has already expressed, via its village chief, the strong desire to optimise the recently provided transportation connections to add value to the unique natural products and increase exports to outside markets, coupled with ecological tourism, to increase the community prosperity. Daxi is, quite literally, 'at the crossroads'.

Daxi is now used to illustrate the possible application of the model, process and planning framework. This process would need to begin by deeper engagement with the Daxi community than was possible during the initial visit by the research team and meeting with the community 'social entrepreneur' chief, to determine 'Where are we now?'. The community context would require further study, relevant stakeholders would need to be identified, and the process of engagement—possibly assisted by a collaboration workshop—would need to draw out stakeholder perceptions and world-views and ascertain their skills, capabilities and resources. After agreeing upon the scope of the exercise, discussion may begin towards accommodating various views under common goals. As described in Chap. 6, the Daxi village chief expressed strong views and a future direction, which we sought to integrate in the transformation statement ('Where we want to get to') of a 'cultured, clean, competitive, confident county'. This would enable villages such as Daxi to develop and extend social enterprises encompassing natural, organic food and medicinal products, ecotourism and farm-stay accommodation, micro-hydro services, in addition to 'eco-system services' and trade in forestry property rights and carbon credits. A series of specific performance indicators could be developed to track progress towards achievement of various goals, such as number of social enterprises created, rise in export income, tourist numbers, jobs created, resource efficiency and the like. Furthermore, it was illustrated that connecting these scenarios and enterprises could uncover synergies for mutual benefit, while utilising and sharing local assets, capabilities, and resources more efficiently and effectively. For example, Jingning food products could be used to market the Jingning brand and attractions to customers, who could be attracted to the county to experience its food and its production modes firsthand, while enjoying the beautiful landscape, clean air water, and learning about the unique *She* culture.

Implementation of such development modes ('How do we get there?') will require widespread community support and may lead to the formation of community cooperatives, which enable the benefits of development to be distributed equitably. Strong business cases will also be necessary to attract external financial support and investment. Government support may also be required, such as by way of policies and regulations to protect the pristine local environment, culture, architecture and the like, which otherwise may be at risk of insensitive tourist development (as in Baisha village) and polluting industries. Partnerships with private corporations (e.g., ecotourism or organic food companies) may also be found to be beneficial, as these could help build local capacity and capabilities as well as providing a source of investment for infrastructure, facilities, value adding eco-industries, and the like. Further partnerships with Chinese and international universities could also lead to training and research in business and management, marketing, hospitality, tourism, and ecological conservation and development which, in turn, may create additional visitor markets in the field of ecological education.

As emphasised at the start of this book, it is critical that communities such as Daxi are empowered to determine their own future, retain leadership of the transformation process, and ownership of their land.

The following discussion highlights other key aspects of the model, process and framework, through examples drawn from further case study communities and from other investigations in China.

8.4.1 Ecological and Equitable Development

In the case of Baisha village, Zhejiang, further advice from architectural and planning experts is required to assist enthusiastic local entrepreneurs construct houses and tourist lodges in harmony with the local landscape and environment, which is threatened by inappropriate and excessive development and tourist waste. The fieldwork revealed other examples of failure of both governments and local communities to recognise the value of their precious culture, heritage and environment.

Similarly, the team witnessed the construction of new villages in Baisha County, Hainan, by external investors; these had entirely replaced the unique structures, despite an effort to 'mimic' the lost, genuine structures and experiences. For example, traditional wooden canoes, which could have added to a rich tourist experience, had been destroyed. The uniqueness and charm had been replaced by a bland

Fig. 8.4 Kitchen of Jingning dwelling

sameness. Such destruction and inappropriate development points to the need for education of villagers, coupled with changes to government policies to protect valuable heritage and the natural environment.

On the other hand, while a Jingning village near the base of a mountain had been abandoned, in accordance with government urbanisation policies, it revealed potential for cultural and ecotourism if sensitive improvements were made to village amenities. The researchers were able to inspect one of the remaining inhabited dwellings, which clearly demonstrated synergies with and respect for the local ecology and environment. As illustrated by Figs. 8.4 and 8.5, the kitchen of the dwelling was open to the mountainside, with a stream—abundant with fish—flowing through the rear of the kitchen. Water was also channelled via bamboo pipes to the nearby crops, and fruit trees and free range chickens provided other natural and organic food.

8.4.2 Partnerships and External Support

The role of professional advisers in assisting communities to transform their mode of development is well-illustrated in the case of Baisha Village, Zhejiang, which shifted from cutting down trees to forest products and ecotourism.

The Service Centre for Modern Science and Technology for Forestry, Lin'an, is a successful example of the same professional adviser engaging with local farming communities to improve their livelihoods, and establishing a cooperative that belongs to the community. Households within the cooperative earn about 100,000

Fig. 8.5 Rear of kitchen showing stream with fish

RMB per annum. The Centre conducts research on ways to improve the productivity of bamboo and soils. For example, bamboo shoots are a valuable food product (about 60 RMB/kg), whilst bamboo can also be used for agricultural and other structures. It is a recognised training centre for national ecological development in China. In addition, the Centre began development of an experimental carbon trading forest. Various types of trees are studied, to achieve ecological balance and carbon emissions reductions. A carbon credits fund is expected to be established, with approval of the China Green Carbon Trade Fund, enabling improvement in household incomes by a further 10,000 RMB per annum.

The Centre is also assisting other regions of China to examine the potential for improvement of degraded lands and gain carbon credits, and to build their local capabilities by training and exchanges. The research team was able to examine such activities of the Centre in Baisha County, Hainan, characterised by high levels of poverty despite a scenic and ecologically rich environment. Baisha is pursuing the path of ecological development, utilising its natural assets for forest products such as tea, bamboo shoots, mushrooms, fruits and traditional Chinese medicines. Of most interest was the experimentation of bamboo plantations on degraded land to deliver multiple benefits, including carbon credits, use of leaves for export as packaging, and the use of the understory for other plant species and for grazing chickens. Similar to the ancient Mangjing tea forests, and approaches in the Dongkeng area, this illustrates pursuit of 'resource productivity', whereby more benefits and services are delivered with less land, materials, energy, water and other associated costs.

Specialised NGOs and business associations also play significant roles in helping rural communities and rural entrepreneurs to gain improved access to much needed financial capital, necessary socio-technical resources, and essential capacity building. For example, China Glory Society (2013) is a Chinese NGO with over 800,000 private businesses as members. The Society is a social organisation for poverty relief, initiated by the Chinese private sector in 1994. It takes development poverty relief as the new modality, poor areas as targets, privately-owned companies as the main players, the communities as actors, and project investment as the main form (Wu, 2012). It has assisted with lifting 130 million people out of poverty through over 20,000 projects and A$10 m investment in rural China. Social entrepreneurs from the Glory Society have formed business partnerships with rural communities for mutual benefit, with a number of successful examples. On the other hand, whilst the 'Brilliant' Tea Company had established a very successful enterprise on Jingmai Mountain, Lancang County, Yunnan, which improved the incomes and education of local *Bulang* employees, the villagers had lost control of their destiny and means of fostering their entrepreneurship, which they were seeking to retrieve.

8.5 Regional Implications and Learnings

As highlighted above, in Chap. 1 and throughout the book, it is most important for communities to retain control of their land and their own development, building self-reliance and self-esteem; choose their own destiny, whilst engaging with others and seeking their support. The following discussion highlights some possible learnings from Australia and Malaysia, where some of the authors have also conducted investigations. Taken together with the findings of the China research, these may have implications for other parts of the region and beyond.

8.5.1 Learning from Australia

Government intervention in Australian indigenous communities, although well-intentioned, have been unsuccessful in many instances, as evidenced by the failed 'kitchen gardens' that were intended to provide local, affordable food for communities in the far north of South Australia (Anangu Lands Paper Tracker, 2011; Martin, 2011). An Australian Government conference on shaping the future of infrastructure in remote indigenous communities, Alice Springs, 5–7 February 2012, conveyed a strong message that indigenous communities prefer to manage their own affairs (Infrastructure Australia, 2012). Indigenous leader Patrick Dodson called for "liberation of Aboriginal people from the public sector", while Patrick Green, chair of *Marra Worra Worra* Aboriginal Corporation, and others pointed out that business cooperatives had been profitable as regional service providers, not merely welfare recipients. Government provision of infrastructure in indigenous communities and monetary handouts had clearly been unsuccessful.

Similarly, as Craig (2007) warned, schemes to enable community capacity building and community development often fail because these imply a lack of capacity by communities: "communities have skills, ideas and capacities; but these are often latent or unacknowledged". Craig asks: "What control do local communities exercise over the capacity-building process? And who defines what a strong community would look like?" (Craig, p. 354).

In Australia, it is recognised that the conventional model based on government handouts, industry donations and charity needs to change. Building cooperative networks with industry and business is seen as crucial for the long-term sustainable development of remote areas, while attracting business investment (Infrastructure Australia, 2011a, 2011b). What are needed are acceptance and mutual realisation that there are common benefits to be gained by business and indigenous communities working together. As long as business interests can be secured, there will be a long-term commitment from both sides to engage in investment towards community development.

On the other hand, Altman (2007) has questioned the dominant indigenous policy approach in Australia, which promulgates a view that indigenous economic development can only be achieved via orthodox engagement with the market either through sale of labour or through operation of commercial businesses. He claims this has been unsuccessful in remote regions, and champions an alternative 'livelihoods approach', referred to as 'the hybrid economy model'. The customary or non-market sector (including hunting, fishing, wildlife harvesting and environmental management of the indigenous estate) has 'a crucially important role to play' in addressing indigenous poverty, in addition to the public (State) and the private (or market) sectors of the economy. Altman observed that indigenous people are poorly remunerated for the provision of a range of environmental services: there are significant opportunities to enhance such indigenous engagement as an element of the hybrid economy. It could be worthwhile to examine the possible extension of such thinking and approaches to ethnic communities in remote areas of China such as the *Bulang* and *She* communities.

8.5.2 Learning from Malaysia

As part of subsequent investigations funded by the Australia-Malaysia Institute, members of the team also examined an 'integrated farm management' and 'smart villages' program conducted in Malaysia by *IRIS Corporation Berhad* in conjunction with the Malaysian Government and rural farming communities (Australia-Malaysia Institute, 2012). Both programs are aimed at raising productivity and livelihoods of farming communities, and improving food security, by means of water saving technologies and ICT, with *Rimbunan Kaseh* being a notable example. However, whilst villagers were able to improve incomes, it appeared that they had largely lost control of their own land and destiny.

This approach contrasted markedly with that of *Cypress Diversified* in Malaysia. The corporation has developed an impressive 'horizontal platform' using ICT to improve rural community livelihoods. This is based upon a database to profile the needs of communities, which is then linked to stakeholders to provide solutions. This 'hands on' program is self-funded by *Cypress*, following the principle of 'by the community, for the community' (POKoK).

Cypress Diversified has developed profiles for approximately 700,000 people in *Pahang* State, Malaysia, where members of the research team conducted fieldwork 17–19 June 2013. This enabled an understanding of the impact of the *POKoK* program on villagers' livelihoods, and included visits to *Kg Taman Sedia village, Cameron Highlands; Kg Orang Asli Sg Ruil, Brinchang* (community displaced by landslide disaster); *Mukim Pulau Tawar, Jerantut; Pemulihan Dalam* Community (rehabilitation centre for young people with learning disabilities); *Ruhah OKU, Maran* (centre for the blind); *Desa Murni* (Rosell) community, *Rompin; Pantai Chedoq, Rompin* beach.

The visits revealed the advantages of the *POKoK* 'community-driven' approach, where 'champions' drawn from the communities provide data, using ICT and 'micro-sourcing', from which community profiles are prepared. Communities are then invited to express their needs and are assisted in elevating these to government and NGOs with the resources to be able to assist communities to develop their own solutions, to run their own enterprises (including those enabled by ICT) and to determine their own future. This could be compared to the approach whereby welfare and facilities are provided to communities, who have a passive, recipient role, as was also evident in some of the communities in Australia.

8.6 Concluding Remarks

The discussion has shown that a multidisciplinary approach to the complex challenges of rural community development can uncover connections and synergies between various topics leading to innovative and breakthrough solutions. For example, interconnected community enterprises based on clean forest products, forest carbon and ecotourism can be underpinned by local infrastructure enterprises (e.g., renewable energy, water, waste management, ICT, transport) and financial mechanisms (e.g., carbon finance), all involving skills development, leadership and social entrepreneurship coupled with partnerships with the corporate sector and investors. Such interconnected approaches are expected to generate increased employment and prosperity, improve social livelihoods, and benefit the environment.

Empowering and assisting communities to operate their own enterprises is seen as fundamental to employment generation, trade, and improving incomes and self-esteem.

Whilst the fieldwork began to engage with the case study communities, time and resources did not permit in-depth studies of the communities and cooperation on strategies for transformation. This is recognised as an area requiring further investigation and, to this end, welcome opportunities for further research collaboration.

References

Altman, J. C. (2007, March). Alleviating poverty in remote indigenous Australia: The role of the hybrid economy. *Development Bulletin, 72*. Retrieved March 3, 2014, from http://caid.ca/AusHybEco020508.pdf

Anangu Lands Paper Tracker. (2011). Mimili: Bush foods enterprise and Anangu employment. Retrieved March 3, 2014, from http://www.papertracker.com.au/index.php?option=com_content&task=view&id=111&Itemid=46

Australia-Malaysia Institute. (2012). *Building bilateral relationships on benefits of ICT for remote communities*. Project by Australian Centre for Asian Business and Malaysia partners. Retrieved March 3, 2014, from http://dfat.gov.au/ami/funded_projects/public_policy.html#ict

China Glory Society. (2013). *China society for promotion of Guangcai program*. Retrieved March 3, 2014, from http://www.cspgp.org.cn/publicfiles/business/htmlfiles/cspgp/s2731/index.html

Craig, G. (2007). Community capacity-building: Something old, something new…? *Critical Social Policy, 27*, 335–358.

Infrastructure Australia. (2011a, October). *Draft strategic framework and funding model*. National strategic policy framework and funding model for infrastructure planning and development in remote indigenous communities.

Infrastructure Australia. (2011b, October). *Infrastructure for sustainable growth: National and international case studies*. National strategic policy framework and funding model for infrastructure planning and development in remote indigenous communities.

Infrastructure Australia. (2012, February). Press release: Indigenous infrastructure conference in Alice Springs, 2. Retrieved March 3, 2014, from http://www.infrastructureaustralia.gov.au/media/2012_02_02.aspx

Martin, S. (2011). Bush garden turns into a wasteland. *The Advertiser, 12*(September).

Ness, D. (2009), Review of conceptual frameworks and available methodologies for the development of eco-efficient and sustainable urban infrastructure, Report for UN Expert Group Meeting 'Developing Eco-efficient and Sustainable Urban Infrastructure in Asia and Latin America', UNESCAP, Bangkok, 10–12 February.

UNESCAP. (2011). *Are we building competitive and liveable cities? Guidelines for developing eco-efficient and socially inclusive infrastructure*. Bangkok, Thailand: Environment and Development Division, United Nations Economic and Social Commission for Asia and the Pacific. Retrieved March 3, 2014, from http://www.unescap.org/esd/environment/infra/documents/Guidelines.pdf

Wu, X. (2012, September 24–28). Integrating development and social public welfare in poverty relief: Roadmap to anti-poverty solutions in ethnic minority regions, International Rural Network (IRN) Forum 2012, Rural and remote resilience: Making the priorities possible, Whyalla, South Australia.

Index

Y. Zhu et al., *Transforming Rural Communities in China and Beyond*,
DOI 10.1007/978-3-319-11319-7

MIX
Papier aus verantwortungsvollen Quellen
Paper from responsible sources
FSC® C105338

If you have any concerns about our products,
you can contact us on
ProductSafety@springernature.com

In case Publisher is established outside the EU,
the EU authorized representative is:
Springer Nature Customer Service Center GmbH
Europaplatz 3, 69115 Heidelberg, Germany

Printed by Libri Plureos GmbH
in Hamburg, Germany